Game Theory and Fisheries

Today, there is a growing sense of urgency among fisheries scientists regarding the management of fish stocks, particularly among those who predict the imminent collapse of the fishing industry due to stock depletion. This book takes a game-theoretic approach to discussing potential solutions to the problem of fish stock depletion. Acknowledging the classification of fish stocks as destructible renewable resources, these essays are concerned with the question of how much of the stock should be consumed today and how much should be left in place for the future.

The book targets both economists and students of economics who are familiar with the tools of their trade but not necessarily familiar with game theory in the context of fisheries management. Importantly, the goal is not to give a summary evaluation of the current views of the "appropriate" response to immediate policy questions, but rather to describe the ways in which the problems at hand can be productively formulated and approached using game theory and couched on real world fisheries.

Game Theory and Fisheries consists of twelve previously published but updated articles in fisheries management, a number of which address a gap in the fisheries literature by modeling and analysing the exploitation of fishery resources in a two-agent fishery, in both cooperative and non-cooperative environments. The author's work ultimately illustrates that the analysis of strategic interaction between those with access to shared fishery resources will be incomplete without the use of game theory.

Ussif Rashid Sumaila is Professor and Director of the Fisheries Centre and the Fisheries Economics Research Unit at the University of British Columbia's Fisheries Centre, Canada. He specializes in bioeconomics, marine ecosystem valuation and the analysis of global issues such as fisheries subsidies; illegal, unreported, and unregulated fishing; and the economics of high and deep sea fisheries.

Routledge explorations in environmental economics
Edited by Nick Hanley
University of Stirling, UK

1. **Greenhouse Economics**
Value and ethics
Clive L. Spash

2. **Oil Wealth and the Fate of Tropical Rainforests**
Sven Wunder

3. **The Economics of Climate Change**
Edited by Anthony D. Owen and Nick Hanley

4. **Alternatives for Environmental Valuation**
Edited by Michael Getzner, Clive Spash and Sigrid Stagl

5. **Environmental Sustainability**
A consumption approach
Raghbendra Jha and K.V. Bhanu Murthy

6. **Cost-Effective Control of Urban Smog**
The significance of the Chicago cap-and-trade approach
Richard F. Kosobud, Houston H. Stokes, Carol D. Tallarico and Brian L. Scott

7. **Ecological Economics and Industrial Ecology**
Jakub Kronenberg

8. **Environmental Economics, Experimental Methods**
Edited by Todd L. Cherry, Stephan Kroll and Jason F. Shogren

9. **Game Theory and Policy Making in Natural Resources and the Environment**
Edited by Ariel Dinar, José Albiac and Joaquín Sánchez-Soriano

10. **Arctic Oil and Gas**
Sustainability at risk?
Edited by Aslaug Mikkelsen and Oluf Langhelle

11. **Agrobiodiversity, Conservation and Economic Development**
Edited by Andreas Kontoleon, Unai Pascual and Melinda Smale

12. **Renewable Energy from Forest Resources in the United States**
Edited by Barry D. Solomon and Valeria A. Luzadis

13. **Modeling Environment-Improving Technological Innovations under Uncertainty**
Alexander A. Golub and Anil Markandya

14. **Economic Analysis of Land Use in Global Climate Change Policy**
    Thomas Hertel, Steven Rose and Richard Tol

15. **Waste and Environmental Policy**
    Massimiliano Mazzanti and Anna Montini

16. **Avoided Deforestation**
    Prospects for mitigating climate change
    Edited by Stefanie Engel and Charles Palmer

17. **The Use of Economic Valuation in Environmental Policy**
    Phoebe Koundouri

18. **Benefits of Environmental Policy**
    Klaus Dieter John and Dirk T.G. Rübbelke

19. **Biotechnology and Agricultural Development**
    Robert Tripp

20. **Economic Growth and Environmental Regulation**
    Tim Swanson and Tun Lin

21. **Environmental Amenities and Regional Economic Development**
    Todd Cherry and Dan Rickman

22. **New Perspectives on Agri-Environmental Policies**
    Stephen J. Goetz and Floor Brouwer

23. **The Cooperation Challenge of Economics and the Protection of Water Supplies**
    A case study of the New York City watershed collaboration
    Joan Hoffman

24. **The Taxation of Petroleum and Minerals**
    Principles, problems and practice
    Philip Daniel, Michael Keen and Charles McPherson

25. **Environmental Efficiency, Innovation and Economic Performance**
    Massimiliano Mazzanti and Anna Montini

26. **Participation in Environmental Organizations**
    Benno Torgler, Maria A. García-Valiñas and Alison Macintyre

27. **Valuation of Regulating Services of Ecosystems**
    Pushpam Kumar and Michael D. Wood

28. **Environmental Policies for Air Pollution and Climate Change in New Europe**
    Caterina De Lucia

29. **Optimal Control of Age-Structured Populations in Economy, Demography and the Environment**
    Raouf Boucekkine, Natali Hritonenko and Yuri Yatsenko

30. **Sustainable Energy**
    Edited by Klaus D. John and Dirk Rubbelke

31. **Preference Data for Environmental Valuation**
Combining revealed and stated approaches
John Whitehead, Tim Haab and Ju-Chin Huang

32. **Ecosystem Services and Global Trade of Natural Resources**
Ecology, economics and policies
Edited by Thomas Koellner

33. **Permit Trading in Different Applications**
Edited by Bernd Hansjürgens, Ralf Antes and Marianne Strunz

34. **The Role of Science for Conservation**
Edited by Matthias Wolff and Mark Gardener

35. **The Future of Helium as a Natural Resource**
Edited by W.J. Nuttall, R. H. Clarke and B.A. Glowacki

36. **The Ethics and Politics of Environmental Cost-Benefit Analysis**
Karine Nyborg

37. **Forests and Development**
Local, national and global issues
Philippe Delacote

38. **The Economics of Biodiversity and Ecosystem Services**
Edited by Shunsuke Managi

39. **Analyzing Global Environmental Issues**
Theoretical and experimental applications and their policy implications
Edited by Ariel Dinar and Amnon Rapoport

40. **Climate Change and the Private Sector**
Scaling up private sector response to climate change
Craig Hart

41. **Game Theory and Fisheries**
Essays on the tragedy of free for all fishing
Ussif Rashid Sumaila

Game Theory and Fisheries

Essays on the tragedy of free for all fishing

Ussif Rashid Sumaila

Routledge
Taylor & Francis Group
LONDON AND NEW YORK

First published 2013
by Routledge
4 Park Square, Milton Park, Abingdon, Oxon OX14 4RN
605 Third Avenue, New York, NY 10017

Routledge is an imprint of the Taylor & Francis Group, an informa business

British Library Cataloguing in Publication Data
A catalogue record for this book is available from the British Library

Library of Congress Cataloging in Publication Data
Sumaila, Ussif Rashid.
Game theory and fisheries : essays on the tragedy of free for all fishing / Ussif Rashid Sumaila.
pages cm
Includes bibliographical references.
1. Fish stock assessment. 2. Fish populations–Mathematical models. 3. Fish populations–Estimates. 4. Game theory. I. Title.
SH329.F56S86 2013
639.3–dc23 2013005787

ISBN: 978-0-415-63869-2 (hbk)
ISBN: 978-0-203-08376-5 (ebk)

Typeset in Times New Roman
by Cenveo Publisher Services

Contents

Figures

Tables

Foreword

The theory of strategic interaction, more popularly known as the theory of games, has over the past few decades come to play an increasingly important role in economics. An indication of this importance is the several Nobel Prizes in Economics awarded to those employing game theory. Game theory has been brought to bear on fisheries economics for just over 30 years. It has been used extensively in the analysis of the economics of the management of international fisheries shared by two or more states (entities). It can be argued that the economic management of such fisheries can be analysed effectively only through the lens of game theory, because strategic interaction between and among states exploiting such shared fishery resources is inescapable.

Game theory is divided into two broad categories: the theory of competitive games and the theory of cooperative games. With reference to international fisheries, theory predicts and practice confirms that competitive fishery games can readily lead to resource destruction. Cooperation does pay, in other than unusual cases.

Professor Sumaila's collection of papers, which makes up this book, demonstrates in a most convincing manner that the potential for the application of game theory in fisheries economics does, in fact, go far beyond the management of internationally shared fish stocks. There are eleven chapters in the book, outside of the introductory chapters. Of these, eight are devoted to fisheries management issues within the coastal state exclusive economic zone (EEZ), such as different interacting fleets exploiting a common resource, and the impacts of marine protected areas. Professor Sumaila shows that the outcomes applying to international fisheries apply equally well to intra-EEZ fisheries. Once again, cooperation pays. His reference to the tragedy of free for all fishing is a reference to competitive fishery games.

Fisheries economists have, heretofore, paid very limited attention to the relevance of game theory to intra-EEZ fisheries management issues. With Professor Sumaila's book now before them, they can afford to do so no longer.

Gordon R. Munro

Preface

Some trace the beginnings of game theory to as far back as 0–500 AD, when The Babylonian Talmud, which contains ancient law and tradition, was put together. But real growth in the application of this branch of mathematics began when John Nash came to Princeton in the early 1950s. In just four papers published between 1950 and 1953, Nash made seminal contributions to both non-cooperative game theory and to bargaining theory. He proved the existence of a strategic equilibrium for non-cooperative games in Nash (1950b), and proposed the "Nash program" in Nash (1951), in which he suggested approaching the study of cooperative games by reducing them to non-cooperative form. He founded axiomatic bargaining theory, proved the existence of the Nash bargaining solution, and provided the first implementation of the Nash program in his two papers on bargaining theory (Nash, 1950a, 1953).

With these four papers, Nash set game theory free by making it more easily applicable in disciplines as wide ranging as philosophy, economics and sustainability of environmental resources. He also helped the profession expand its focus beyond non-cooperative games (mainly zero sum – your loss is my gain and vice versa) to include cooperative games. The latter contribution is huge because it showed that cooperation is possible in competitive situations, and that all players can gain by engaging in cooperative behavior. For these contributions, John Nash was honored with a Nobel Prize in economics in 1994, and was immortalized with the famous film *A Beautiful Mind*.

The application of Nash's concepts of non-cooperative and cooperative game theory in the fisheries economics literature started with Munro (1979). Since the publication of this paper, several game-theoretic papers have appeared in the literature, developing models to analyse internationally shared fish stocks (see Bailey *et al.*, 2010 and Hannesson, 2011 for recent complementary reviews). The main contribution of this book is that it demonstrates, with several examples, that cooperative and non-cooperative game theory are equally valuable for analysing domestic fisheries, i.e. those based on fish stocks that do not straddle (extend) into international waters or even the exclusive economic zones of other maritime countries (Sumaila, 2012). In all the cases presented, it is shown that "free for all fishing," i.e. non-cooperation, is bad for both the fish and the fishers, while cooperative management works both in terms of sustaining the fish stock and providing much larger economic benefits to cooperating players.

Ussif Rashid Sumaila

Acknowledgements

I must first thank my co-authors for some of the papers included in this book. After the publication of the first set of papers contained in the book, I got into fruitful collaboration with my colleague and friend, Claire Armstrong of the University of Tromsø, during which we produced a series of papers that built on the methodology I developed to analyse a number of issues such as the efficiency and distributional effects of implementing marine protected areas, and the effects of cannibalism on the economics of a shared fish stock. I wish to thank Claire for several years of excellent collaboration.

Megan Bailey, my former student, applied game theory to fisheries in her doctoral research and was also co-author of one of the papers included in the book. Megan was in fact the first one to work on this book as a research assistant. I thank Megan for many years of working together on game theory and for her contribution to making this book a reality. Similarly, my former postdoc, Ling Huang, co-authored the paper that is the basis of Chapter 12 of this book. My best wishes to Megan, who is currently a postdoc fellow at Wageningen University, and Ling, currently assistant professor at the University of Connecticut, as they develop their careers.

I am very grateful to my lead PhD supervisor (Sjur Didrik Flåm) and co-supervisor (Rögnvaldur Hannesson) for introducing me to game theory and fisheries. I was very fortunate to have such a productive duo as supervisors as I quickly learned the value of "getting things done." The combination of having a solid applied mathematician (Flåm) and an accomplished natural-fisheries-resource economist (Hannesson) as supervisors also showed me, early in my career, the value of interdisciplinarity in studying real world issues, such as how to sustainably manage and use fishery resources for the benefit of all generations.

Gordon Munro has also played an important role in my career development. First, he was the one who truly started the study of fisheries using game theory with his 1979 seminal paper (Munro, 1979). Second, I came to UBC in 1995 as a visiting scholar because of Gordon, as I was keen to work with the man who started it all. Third, it was Gordon who introduced me to the Fisheries Centre, and this happened by chance. He could not find me office space at the Department of Economics, so as a Faculty Associate of the Fisheries Centre, he contacted the then FC Director, Dr Tony Pitcher, and got a desk for me at the Centre, and the rest of my relationship with the FC is now history. Thank you, Gordon.

I thank Daniel Pauly for two reasons. First, without his encouragement this book would not have been written. He worked hard to convince me that doing this book would be truly worth the effort. Second, Daniel Pauly painstakingly reviewed the final draft of the book and thereby helped to improve it significantly.

I need to thank colleagues and students at the Fisheries Centre for stimulating interactions since my arrival at the Centre. I still remember the first talk I gave at the Centre and the stimulating comments I received – thanks to all of you.

Carmel Ohman needs special mention for providing me with essential research assistantship during the final stage of compiling the book: being an English major certainly helped. I also thank Rachel 'Aque' Atanacio for redrafting many of the figures in the book.

I also want to take this opportunity to thank the publishers of the original articles that granted me copyrights so I could republish figures that first appeared in their journals.

Over the years, I've been supported and funded by several organizations but key sponsors are the Research Council of Norway, the *Sea Around Us*, the Pew Charitable Trusts, the Kingfisher Foundation, Conservation International, the World Wildlife Fund, and the Social Sciences and Humanities Research Council of Canada. I am grateful for the support.

Finally, I have to thank Mariam Sumaila, who, like Daniel Pauly, made me not only to see the need to do a book like this but also made sure that I could not avoid writing it. Essentially, they ensured that even though I tried to run from writing the book, I could not hide – I thank both of them immensely for helping me get this done!

Symbols and acronyms

α	Constant parameter in the Beverton–Holt recruitment function (see γ)
γ	Constant parameter in the Beverton–Holt recruitment function (see α)
δ	Discount factor
π	Single period profit
Π	Discounted sum of single period profits
$\Pi_i\ K_T\ x\ K_C \rightarrow \Re$	where K_T and K_C are the pure strategy sets of player $i =$ T, C, that is, the set of fishing capacity
ψ	Fishing costs per vessel, which consist of fixed costs (ϕ_i) and variable costs (ξ_i)
ψ_n	Net migration of cod from the protected to the unprotected area
a	Age group of fish
α	ANCS North-east Atlantic cod stock
B	Spawning biomass in weight
BFT	Atlantic bluefin tuna
C	Coastal Fisheries Management Agency
$C(.)$	Catch cost function
CITES	Convention on International Trade in Endangered Species of Wild Fauna and Flora
CPC	Contracting Parties & Cooperating non-Contracting Parties, Entity and Fishing Entity
CPs	Contracting Parties
CPUE	Catch per unit effort
DAP	Dedicated Access Privileges program
δ	Fishing effort
EC	European Community
EEZ	Exclusive economic zone
EFF	European Fisheries Fund
EU	European Union
$f(.)$	Beverton–Holt recruitment function

FADs	Fish aggregating devices
FIFG	Financial Instrument for Fisheries Guidance
G	Natural growth function of a sub-stock
h	Catch of fish
i	Players in a game; where player i only catches sub-stock i, this also represents "owners" of a sub-stock
ICCAT	International Commission for the Conservation of Atlantic Tunas
ICES	International Council for the Exploration of the Seas
ICSEAF	International Commission for the South-East Atlantic Fisheries
IMR	Institute of Marine Research, Bergen, Norway
ITQ	Individual transferable quota
IUU	Illegal, unreported and unregulated fishing
$k/(1+\omega) \approx k$	Cost of engaging one fishing fleet for one year; ω is a cost parameter
LLD	Owners of deepwater longlines
LLS	Owners of shallow longlines
M	Discounted single period profit of a sub-stock
MENA	Middle East and North Africa
MEY	Maximum economic yield
MFMR	Ministry of Fisheries and Marine Resources, Namibia
MPA	Marine protected areas
MSY	Maximum sustainable yield
n	Denotes stock size as number of fish
N$	Namibian dollar
NE	Nash equilibrium
NOK	Norwegian kroner
NMFS	National Marine Fisheries Service
NTAC	Norwegian share of the total allowable catch
OA	Open access
OAS	Open access plus subsidy
p	Proportion of mature fish
PMR	Protected marine reserve
prof_{com}	Objective function
PS	Owners of purse seines
PV	Present value
Q	Quota
qs	Quota share
q	Catchability coefficient, that is, the share of fish biomass being caught by one unit of fishing effort
r	Discount rate
rev	Revenue from fishing
RFMO	Regional Fisheries Management Organization

R&M	Reporting and Monitoring
s	Natural survival rate of fish
SBT	Southern bluefin tuna
SCRS	Standing Committee on Research and Statistics
SSB	Spawning stock biomass
t	Fishing period in years
T	Trawl Fisheries Management Agency
TAC	Total allowable catch
v	Price per kilogram of fish
w	Weight of fish
WCPO	Western Central Pacific Ocean
w_{sa}	Weight at spawning
WWF	World Wildlife Fund
x	Biomass of a sub-stock

1 Introduction

The essays in this book illustrate what I call the "tragedy of free for all fishing." As most readers of this book may be aware, Hardin (1968) made the term "tragedy of the commons" popular, but a close reading of his seminal paper reveals that what he was actually concerned about is the "tragedy of free for all," where a commonly shared resource is accessed by many without effective regulation (e.g. Hawkshaw *et al.*, 2012). The essays in this book demonstrate, through the application of game theory, that even fish stocks that are owned commonly by a number of agents (e.g. two countries) can be managed to avoid the "tragedy of the commons" by putting in place an effective cooperative management regime – a point that was demonstrated by Nobel Laureate Ostrom (1990) and her collaborators more generally.

Abstracting from the complications normally encountered in attempting to give a precise and concise definition of "natural resources," a resource is defined as a natural resource if it has the following features (Mclnerney, 1981):

- The maximum stock of the resource that could be utilized is totally fixed, having been predetermined before humans commenced any economic activity; or
- To the extent that the available stock changes, it does so at a "natural" biological or biochemical rate.

The first feature is shared by resources known collectively as "non-renewable resources." Examples of these include fossil fuel, metal ores, and land area. Renewable resources such as fish stocks and forestry resources, share the second feature. The focus of this book is on the application of game theory to the management of renewable natural resources with particular emphasis on fisheries.

Fish belong to the class of natural resources that may be classified as destructible renewable stock resources (Mclnerney, 1981). They therefore feature the following characteristics:

- "utilization" of a unit of the fish resource implies its destruction. That is, the unit is completely and irrevocable lost; and

- the fish stock can be augmented again to enable a continuing availability through time.

Thus, fish (as for other renewable natural resources), have the special feature that, even though its utilization results in depletion, new stocks are created by a process of self-generation. The regeneration occurs at a "natural" or biological rate, often directly dependent upon the amount of original stock remaining unutilized. The essence of fishery economics stems from the stock characteristics of fisheries and the fact that the rate of biomass adjustment of a fish stock is assumed to be strictly a function of that stock (Tomkins and Butlin, 1975). Essentially, the central problem of natural resource economics, at large, and fisheries economics in particular, is intertemporal allocation. In other words, natural resource economists are mainly concerned with the question of how much of a stock should be designated for consumption today and how much should be left in place for the future. This central problem, together with its many extensions and varied forms, has been a "center of attraction" for many economic studies, debates, and discussion in the literature. This book will review, in particular, the game theoretic approach to the solution of this problem, together with all its ramifications.

The book is meant for students and scholars of fisheries science, economics, and management who are familiar with the tools of their trade, but not necessarily familiar with the game theoretic approach to the management of natural resources such as fish stocks. The ways in which the problems at hand can be productively formulated and approached are described using game theory. Rather than attempting to trace each application through its long and complicated history, the main problems addressed and the main results derived thereof are presented.

The book contains 12 chapters, based mostly on previously published articles. This collection of essays, together, provides a window into the many applications of the theory of games, whose theoretical foundations were laid by Nobel laureate John F. Nash of *A Beautiful Mind* fame (Nasar, 2002). The application of this theory to fisheries management started in earnest in the late 1970s with a paper by G. Munro (1979). The contribution of this book is that it demonstrates how game theory can be used to analyse shared stocks within country exclusive economic zones (EEZs). This is important because most applications of game theory to fisheries are couched on transboundary or highly migratory stocks.

The book has a broad coverage in terms of the types of fisheries, the geographical scope of its coverage, and the types of externalities being analysed. It features fisheries from around the world, including applications of game theory to the Namibian hake (cape hake, *Merluccius capensis* and deep-water hake *Merluccius paradoxus*) fishery; the cod (*Gadus morhua*) and capelin (*Mallotus villosus*) fisheries of the Barents Sea; and the tuna (bigeye *Thunnus obesus*, yellowfin *Thunnus albacares,* and skipjack *Katsuwonus pelamis*) fisheries of the Western Central Pacific Ocean. The book also explores the economic effects of dynamic and species interaction externality; implementing marine reserves;

irreversibility in fishing capital investment; cannibalistic tendencies of certain species; and the implementation of individual transferable quotas.

Chapter 2 provides a review of the application of game theory to the management of fishery resources that are shared by more than one agent (individual, country, region, fishing group, etc.). The key conclusion that may be drawn from the essay is that the analysis of strategic interaction between fishers who have access to shared fishery resources would be incomplete without the use of game theory.

In Chapter 3, a two-agent model that assumes perfect malleability of fishing capital for the exploitation of the North-east Atlantic cod stock is developed to investigate the economic benefits that can be realized from the resource, and the effect of exploitation on stock sustainability under cooperation and non-cooperation. The two agents are identified in this chapter as a trawl fishery and a coastal fishery. Here, conflicts between agents arise mainly from the differences in fishing gear and grounds, and the age group of cod targeted by the two agents. It is shown that given available data, the best outcome is obtained under cooperation with side payments and no predetermined catch shares, in which case the coastal fishery buys out the trawl fishery. However, sensitivity analysis shows that if the price premium assumed for mature cod is taken away, the trawl fishery takes over as the producer of the best outcome for players.

The theme of Chapter 3 is pursued further in Chapter 4. A similar two-stage, two-player non-cooperative game model is developed under a non-malleable or irreversible capital investment assumption. The goal here is to predict the number of vessels that players in such a game will find in their best interest to employ in the exploitation of the North-east Atlantic cod stock in the Barents Sea, given a non-cooperative environment and the fact that all players are jointly constrained by the population dynamics of the resource. The predictions obtained are then compared with (i) the sole owner's optimal capacity investments for the two players; (ii) the results in Chapter 3, where perfect malleability of capacity is assumed implicitly; and (iii) available data on the North-east Atlantic cod fishery.

Chapter 5 develops a bioeconomic model for two Barents Sea fisheries that attempts to capture the predator–prey relationships between cod and capelin, the two main species in the ecosystem. The aim is to analyse joint (cooperative) versus separate (non-cooperative) management of this predator–prey system with a view to isolating the efficiency loss due to separate management. Using a game theoretic framework and a multi-cohort age-structured bioeconomic model, joint and separate management equilibrium outcomes are computed to help investigate the effects of changes in economic parameters on the computed results. In this way, the economic consequences of the predator–prey relationships between cod and capelin, and the externalities due to non-cooperation are explored. Results of the study suggest that (i) under the prevailing market conditions, it is economically optimal to exploit both species (rather than just one of them) under joint management; (ii) in comparison with the separate management outcome, a severe reduction of the capelin fishery is called for under joint management; and (iii) the loss in discounted resource rent resulting from the externalities due to the natural interactions between the species is significant, reaching up to almost a quarter of what is achievable under separate management. The work in this chapter is in a way a precursor to the more recent works on the role of forage fish in marine ecosystems (e.g. Hannesson *et al.*, 2009; Herrick *et al.*, 2009; Pikitch *et al.*, 2012).

In Chapter 6, intra-stock relations such as cannibalism and growth enhancement are investigated to determine the economically optimal sharing of a fish resource between heterogeneous fishing agents. The sharing of resources between different vessel groups is often left for political decision-making. Nonetheless, such decisions may have both biological and economic consequences. This becomes quite clear when different fishing groups exploit different sections (age groups) of a stock that has intra-stock interactions in the form of cannibalism. A two-agent bioeconomic model with cannibalism is developed and used to determine (i) total allowable catches (TACs) for cod; and (ii) the optimal proportion of the TAC that should be caught by the different vessel groups in the fishery. Applying biological and economic data in a numerical procedure, and comparing the results obtained to previous studies, it is shown that intra-stock interactions, such as the presence of cannibalism, have a significant impact on who should take what proportion of the TAC, and hence, the standing stock size and discounted resource rent achievable. In contrast to other studies, it is found that the optimal catch requires that both trawlers and coastal vessels catch the fish resource. In addition, the results indicate that, from a bioeconomic perspective, the existing trawler fleet's catch share in the cod fishery is too high.

Chapter 7 has two goals. First, the allocation rule, applied to split the Norwegian total allowable catch for cod between coastal and trawler vessels, is studied. Second, the bioeconomic implications of an individual transferable quota (ITQ) management system for this fishery is analysed. A cannibalistic bioeconomic model with cooperative game theory is developed. Key results from the study are (i) the current allocation rule acts in opposite fashion to what may be considered bioeconomically optimal; and (ii) an ITQ system for this fishery is likely to result in economic losses, as the biological advantages of fishing with the two vessels types may be lost.

What bioeconomic benefits can be expected from the implementation of marine protected areas (MPAs) in a fishery facing a shock in the form of recruitment failure, and managed jointly compared to separately? What are the optimal sizes of MPAs under cooperation and non-cooperation? These are the questions explored in Chapter 8. A computational two-agent model is developed, which incorporates MPAs using the North-east Atlantic cod fishery as an example. Results from the study indicate that MPAs can protect the discounted resource rent from the fishery if the habitat is likely to face a shock, and fishers have a high discount rate. The total standing biomass increases with increasing MPA size but only up to a point. Based on the specifics of the model, the study also shows that the economically optimal size of MPA for cod varies between 50% and 70%, depending on (i) the exchange rate between the protected and unprotected areas of the habitat; (ii) whether fishers behave cooperatively or non-cooperatively; and (iii) the severity of the shock that the ecosystem may face.

Chapter 9 studies the distributional and efficiency effects of MPAs using the North-east Atlantic cod stock as an example. A model with two players targeting different age groups of cod is developed to examine how protected areas may affect payoffs to the players under cooperation and non-cooperation. It is found that depending on the *ex ante* and *ex post* management regime, win-win, lose-lose, or win-lose outcomes may emerge with the implementation of MPAs. When the *ex post* management is cooperation, both players gain, while *ex post* non-cooperative behavior results in gains only to one of the players.

A sequential game theoretic model involving the purse seine fleet used by domestic countries of the Western Central Pacific Ocean (WCPO) tuna stock, such as the Philippines and Indonesia, and the longline fleet used mainly by distant water fishing nations to target tuna in the same region is developed in Chapter 10. Purse seines target mainly skipjack but in so doing they also catch a sizable quantity of juvenile bigeye and yellowfin tuna. The longline fleet is split into two groups, that is, the shallow water longline fleet that targets both bigeye and yellowfin, and the deep water longline fleet, which targets mainly bigeye stocks. The purse seine fleet takes juvenile bigeye and yellowfin tuna before the longline fleet gets the chance to target them, thereby creating a sequential game situation. Joint (cooperative) versus separate (non-cooperative) management of these three stocks of tuna in the WCPO are developed, with a view to isolating the net benefit loss due to separate management. Results of the analyses suggest that (i) it is economically optimal to cut back significantly on the bycatch of bigeye and yellowfin by reducing the use of fish aggregating devices; and (ii) such a cut in bycatch will result in a loss to the domestic countries that target skipjack but this loss is much smaller than the gain in the potential benefit to the longline fleet. For joint management to be implemented, an institutional arrangement is needed to allow domestic countries using purse seines to share in the gains from cooperation.

Chapter 11 presents a model for Namibian hake, which incorporates the biology, gear selectivity and the economics of the hake fisheries in a framework that allows the analysis of fishing gear impacts on the potential economic gains from the resource. The objective is to produce quantitative results on the key variables of the fishery – namely, resource rent, standing biomass, and catch levels – that will support the optimal sustainable management of one of Namibia's most valuable fishery resources. Outcomes for three management scenarios are produced, (i) command; (ii) cooperative; and (iii) non-cooperative. For each of these, results are presented for two different assumptions of the economic setting under which the managers of the fishery operate, that is, a fully economic setting and a setting with cost-less labor inputs. As would be expected, different management scenarios and assumptions about the economic setting impact on the results derived from the model in significant ways.

An explanation of why the attempt to manage Atlantic bluefin tuna (*Thunnus thynnus*) stocks in the Mediterranean Sea has so far failed is given in Chapter 12.

Stock status of the fish is reviewed, and resource rent estimated, to evaluate the fishery's management system. It is determined that the non-restrictive implementation of International Commission for the Conservation of Atlantic Tunas (ICCAT) policies is the institutional reason for its management failure, while the common-property and shared stock nature of this fishery is the fundamental reason. To address these major issues, policy schemes that can help ensure sustainable management of this valuable fish stock are proposed.

2 Game-theoretic models of fishing[1]

Introduction

The solution to the central problem of *intertemporal allocation* has been elusive for the following reasons. First, renewable natural resources, such as fish stocks, are often "common property," in which several entities have fishing rights to the resource (e.g. Sumaila, 2012a). In particular, certain fisheries are transboundary and/or straddling in nature.[2] Second, some species of fish are long-lived, such that whether juveniles or mature fish are caught can have important biological and economic consequences. Third, in multispecies systems, there is usually some form of natural interaction between species, which has both biological and economic consequences. Fourth, different vessel types employed in the exploitation of the resource have different effects on the health of the stock, and the economics of the fishery. Fifth, capital embodied in the exploitation of natural resources is often non-malleable, which can have an impact on management plans. Sixth, there is the problem of uncertainty about the biology and economics of the resource. Seventh, the problem of market interaction in both factors and products must be dealt with. A related topical issue is how global warming is likely to complicate fisheries management (Sumaila *et al.*, 2011). As demonstrated in the sections that follow, the fisheries economics literature is rich in attempts to address problems of *intertemporal allocation*, except the challenge posed by global warming where work is beginning to trickle in (e.g. Miller and Munro, 2004).

Models of fishing

Open access and sole ownership fishery models

Economists have traced the main problem of the fishing industry to its unique "common property" characteristics (Copes, 1981). The first comprehensive analysis of this problem was by Gordon (1954) (see also Hannesson, 1993a; Clark, 2010; Bjørndal and Munro, 2012). The common property characteristic of the fishery is necessarily associated with both open access and the lack of delineated rights to the fishery (see Bjørndal, 1992, for a review of the social planner (sole ownership) and the open access equilibrium outcomes). Earlier published analyses of fisheries economics (Christy and Scott, 1965; Smith, 1969) have been concerned with two contrasting systems of access rights: (i) full rights and (ii) no rights. These two systems yield unique "Nash non-cooperative outcomes,"[3] namely, the sole ownership (social planners') outcome for the former, and the open access outcome for the latter. The open access or the "tragedy of the commons" outcome (Hardin, 1968; Hawkshaw *et al.*, 2012) is easy to implement but most wasteful. A solid theoretic discussion of this outcome is given in Clark and Munro (1975) and Clark (1990). The social planners' outcome, by reducing play to a sole owner, is almost impossible to realize in practice because of the constant threat of new entrants into the fishery. The sole ownership equilibrium, however, has excellent efficiency properties. It is usually used as a reference point for the analysis of real world situations.

Game-theoretic models

Game theory is a mathematical tool for analysing *strategic interaction*. For example, suppose a few firms dominate a market, or a few groups of individuals or entities have fishing rights to a common property resource, or countries have to make an agreement on trade or environmental policy. Each agent in question has to consider the other's reactions and expectations regarding their own decisions.

With the development of game theory[4] came its use for analysing problems not only in economics but also in such diverse areas as political science, philosophy, and military strategy.[5] Currently, there is an explosion in the use of game theory and applications thereof in virtually all areas of economics (e.g. Madani and Diner, 2012, on water management).

Game-theoretic fisheries models are made up of a combination of a biological model of fisheries and one of the solution concepts of Nash, or their refinements. The biological models underlying such game-theoretic models can be classified into two main categories (Reed, 1980). First, models of the *lumped parameter* type, for which the models of Ricker (1954) in discrete time, and of Schaefer (1957) in continuous time, are the most widely used. Second, the so-called *cohort models*, which explicitly recognize that fish grow with time and suffer natural mortality. The most commonly used model in this class is that of Beverton and Holt (1957). Reed (1980) argues that both the age at which fish are captured and the relationship between parent stock and recruitment play an important role in determining yields in many commercially important fisheries. Therefore, it would seem reasonable to consider optimal catching using a model which incorporates both a cohort structure and dependency of recruitment upon the parent stock. One model with both of these characteristics is the Leslie matrix model (Lewis, 1942; Leslie, 1945).[6,7]

Cooperative and non-cooperative management

Nash (1953) was the first to explicitly distinguish between cooperative and non-cooperative games. He classified games in which binding agreements are not feasible to be non-cooperative, and those in which binding agreements are feasible, cooperative games. Both of these types of games have been used to analyse the exploitation of fishery resources (Bailey *et al.*, 2010). Usually, models are developed to study what happens both to the biology and economics of a fishery under cooperation and non-cooperation, with the aim of isolating the negative effects of non-cooperation (Levhari and Mirman, 1980; Fischer and Mirman, 1996; Mackinson *et al.*, 1997; Lindroos, 2004).

In undertaking a cooperative management analysis, Munro (1979) combined the standard economic model of a fishery with cooperative game theory. It is shown in this study that if the cooperative management is unconstrained, i.e. if allowances are made for time-variant catch shares and for transfer payments, then to achieve optimal joint catch demands that the patient player should buy out its impatient partner entirely at the commencement of the program and manage the resource as a single owner (Munro, 1991a). Thus, achieving what Munro calls an *optimum optimorum*.[8] Chapter 4 develops an applied computational game-theoretic model in which two vessel types are organized as separate agents, who exploit a shared stock (the North-east Atlantic cod stock). The results of this study confirm the main theoretical finding of Munro (1979).

The analysis of cooperative non-binding programs is more difficult (Munro, 1991a). The key to the solution of such programs is for each player in the game to devise a set of "credible threats" (Kaitala, 1985). Kaitala and Pohjola (1988) provide a good example of non-binding cooperative management. In their model, the management program is modeled as a differential game in which memory strategies are used. Vislie (1987) developed a simplified version of Munro (1979), which he used to derive a self-enforcing sharing agreement for exploiting transboundary renewable resources in cooperation without strictly (judicially) binding contracts.

Krawczyk and Tolwinski (1991) consider a feedback solution to an optimal control problem with nine control variables for the southern bluefin tuna (SBT). Kennedy and Watkins (1986), instead, consider a cooperative solution for the SBT management problem modeled as a two-agent, optimal control problem with linear dynamics. Both papers use multi-cohort biomodels to determine optimal time dependent quotas. To solve their models both studies employ the perturbation method developed in Horwood and Whittle (1986). Using a Nash co-operative game, Klieve and MacAulay (1993) show the importance of fishing strategies for SBT that take into account the age distribution of the catch. If Japan and Australia act according to a cooperative game, the optimal fishing strategy would involve Australia avoiding the fishing of very young cohorts and Japan taking a moderate catch in subsequent older age classes but not the oldest of the age classes.

Dynamic externality

Dynamic externality is the bioeconomic loss which arises when a dynamic population is exploited by a finite number of fishers. Levhari and Mirman (1980) study this kind of externality by using the concept of Cournot–Nash equilibria. Clark (1980) considered a limited access fishery as an N-person, nonzero-sum differential game. The work reported in this book uses computational game-theoretic models of fishing that study the consequences of dynamic externality. All these papers show that, no matter the details of the models developed, the negative bioeconomic effects of dynamic externality are quite significant.

Market externality

Market externality occurs when market-clearing prices depend on the catch of fish, and therefore the quantity supplied, thereby generating an externality in the sense that if one agent supplies more fish the payoffs of other agents are affected. Dockner *et al.* (1989) presented a generalized Gordon–Schaefer fishery model to a duopoly. The main difference between this model and "no-market" interaction models, such as in Clark (1980) and those reported in this book, is that it is an oligopolistic model rather than a competitive one.[9] It assumes that the price of landed fish is not constant, but depends on the quantity caught by all producers, implying that the interaction at the marketplace, while not the only interaction between agents, is important. The paper studies the impact of different oligopoly strategies, i.e. Nash and Stackelberg, on prices, quantities, and payoffs to the players. The authors set up a non-cooperative game which they solve both analytically and numerically by using the equilibrium concepts of Nash and Stackelberg. Their analysis shows that in both the Nash and Stackelberg cases, the player with the smaller unit cost is able to choose higher catch rates than his or her opponent. They also find that the game is Stackelberg dominant. This means that the payoffs to both players are higher in the Stackelberg case than in the corresponding Nash case. Another finding of theirs is that in the Stackelberg case, any information disadvantage in the sense of Stackelberg followership can be eliminated by a more efficient technology. Datta and Mirman (1999) found a sub-game perfect Cournot–Nash equilibrium in a study of the conditions under which such an equilibrium may be efficient. The authors' goal was to analyse the role of different externalities in generating economic inefficiency.

Multispecies interaction externality

Quirk and Smith (1977) and Anderson (1975a, 1975b) were among the first theoretical papers to appear in the fisheries economics literature on ecologically interdependent fisheries. Both study and compare the free access equilibria and the social optima in such systems. They derive necessary conditions for optima and interpret these in general terms. Hannesson (1983) extends the results of these two papers to address broader questions such as, is there a price at which it is economically sensible to switch from exploiting the prey to exploiting the predator in such systems?

Fischer and Mirman (1992, 1996) and Flaaten and Armstrong (1991) are theoretical papers which analyse interdependent renewable resources using game-theoretic models. These papers assume single cohort growth rules to derive general theoretical results. Some of the studies reported in this book are empirical studies of the Barents Sea fisheries, which explicitly recognize that fish grow with time and that the age groups of fish are important both biologically and economically. For another study of such problems in a strategic context see Clemhout and Wan (1985).

Transboundary/migratory/straddling stock models

One can distinguish between three types of transboundary fishery resources. First, fish stocks that migrate between the exclusive economic zone (EEZ) of two or more coastal states, which may be considered transboundary resources "proper." Second, highly migratory stocks, which in effect refer to tuna. Third, the so-called "straddling" fish stocks, i.e. those stocks that migrate between the EEZ of one or more coastal states and the high seas (Munro, 1996).

Analysis of the management of transboundary resources "proper" is treated in Munro (1990), McRae and Munro (1989) and Munro (1991a). Flaaten and Armstrong (1991) and Flaaten (1988) are treatments of transboundary fishery problems involving Norway and the former Soviet Union.[10] Recent contributions to the area of migratory fisheries are: Munro (1991b), Arnason (1991) and Fischer and Mirman (1992, 1996). It is demonstrated in Munro (1979) and Levhari and Mirman (1980) that, whatever the scenario chosen, the outcome to the fishing nations of non-cooperation is of unquestioned undesirability (Munro, 1991b). This is because the outcome is simply Pareto-inefficient, implying that the payoff to some of the players can be increased without necessarily decreasing those of others.

The theory of transboundary fishery resources has been used in the context of different user groups and/or vessel types exploiting a shared stock. Munro (1979) and Sumaila (1995) are examples in which studies of the exploitation of a shared stock are organized around the vessel types employed in the exploitation of the resource.

Recent conflicts, such as those between Canada and the EU over stocks straddling between Canada's EEZ and the high seas, have generated interest among fisheries economists on the management of straddling fish stocks, with Kaitala and Munro (1993, 1995) leading research efforts. Their work has thus far shown that the non-cooperative theory developed for the study of transboundary resources also applies to straddling stocks. This is, however, not the case when it comes to cooperative theory. Here, the cooperative theory of transboundary resources breaks down because of the so-called "entry-exit" problem implied by the "Draft Agreement for the Implementation of the Provisions of the United Nations Convention on the Law of the Seas of 10 December 1982 Relating to the Conservation and Management of Straddling Fish Stocks and Highly Migratory Stocks" (1994).

Malleable and non-malleable capital models

A number of papers have appeared in the fishery economics literature that focus, in part, on the irreversibility of capital employed in the exploitation of fishery resources. Examples include Clark and Kirkwood (1979), Clark *et al.* (1979), Dudley and Waugh (1980), Charles (1983) and Charles and Munro (1985). Among these examples, only Dudley and Waugh (1980) consider, qualitatively, investment decisions in a fishery with more than a single agent. Chapter 4 of this book provides a quantitative analysis of a two-agent fishery where the irreversibility of capital is the central assumption. The negative bioeconomic effects of irreversibility of capital were shown to be significant.

Fisheries management models with uncertainty

Uncertainty is certainly an obstacle for sustainable fisheries management, the main sources of which include: firstly, the dynamic nature of fish populations in the wild and the variability and complexity of the marine ecosystems of which they are a part, and secondly, the impact of fishing activity upon the resources, and the fact that perfect monitoring and control of catching in marine capture fisheries will forever be problematic.

Uncertainty has been classified into two broad categories (Sumaila, 1998a). First degree uncertainty consists of "random effects whose future frequency of occurrence can be determined from past experience" (Walters and Hilborn, 1978). Hence, it is possible to construct objective probability distributions to capture this class of uncertainty. Second degree uncertainty, usually termed "true uncertainty," covers events that cannot be predicted, and for which objective probability cannot be estimated (Sumaila, 1998a). It is possible to reduce this class of uncertainty through further research but to eliminate it completely is but a dream: an irreducible level of uncertainty will always exist.

To date, most stochastic economic models of fisheries incorporate only first degree uncertainty Andersen and Sutinen (1984). Protected marine reserves (PMRs) have been advanced as a viable tool for dealing with second degree uncertainty. A key effort in this direction is the work of Lauck *et al.* (1998). This paper has explicitly linked the mitigation of second degree or true uncertainty to the creation of PMRs. Many biological papers have promoted the establishment of PMRs as a viable alternative where other forms of fisheries management are impracticable or unsuccessful (Wallis, 1971; Davis, 1981; Bohnsack, 1990). It remains to be seen what bioeconomic models of marine reserves will demonstrate about the use of marine reserves to hedge against uncertainty.[11]

Computational methods

The key to the empirical applications, in fisheries economics, of the theoretical assertions of game theory is the development of computational techniques for identifying the predicted equilibrium solutions. Three types of equilibrium concepts or informational assumptions are used in game-theoretic models: *open-loop, feedback*, and *closed-loop*. With open-loop information in dynamic games, players cannot observe the state of the system after time equals zero. Even if they can, it may not be possible for them to do anything about it. In other words, they can commit to their controls only at the start of the game. Feedback and closed-loop are rules for choosing controls as functions of the state (stock). The difference between the two information structures is that for feedback controls, which are Markovian in nature, players know only the current state (i.e. the payoff relevant actual information), whereas closed-loop information includes the way in which the stock has evolved so far in the game (Slade, 1995). Feedback and closed-loop controls allow the payer more rationality and flexibility but due to the difficulty of computing these solutions, there has been a tendency in the literature to resort to the use of open-loop solution concepts.[12] There are other reasons for the continued use of the open-loop equilibrium concept in the literature. In the first place, more rationality and flexibility does not necessarily mean that closed-loop solutions are always better than their open-loop counterparts. In the discussion of rules, or open-loop in our context, versus discretion, or closed-loop in the macroeconomics literature, rules are shown to often produce more desirable outcomes than discretion (Kyndland and Prescott, 1977). Second, the open-loop solution concept can be used with a more complex information structure, known as *piecewise deterministic games* (Haurie and Roche, 1993).

Many algorithms for the computation of economic equilibria have been presented in the computational literature (Bertsekas and Tsitsiklis, 1989). Examples of methods for computing game-theoretic equilibrium solutions are: the perturbation method of Horwood and Whittle (1986); the methods used to construct and estimate game-theoretic models of oligopolistic interaction (Slade, 1995); methods for computing cooperative equilibria in discounted stochastic sequential games (Haurie and Tolwinski, 1990); and algorithms from non-smooth convex optimization, in particular, subgradient projection and proximal-point procedures (Cavazutti and Flåm, 1992). The latter class of algorithms are intuitive because they are "behavioristic," modeling out-of-equilibrium behavior as a "gradient" system driven by natural incentives.

Concluding remarks

In terms of policy, this chapter shows that results derived from game-theoretic models of fishing have produced insights that have been beneficial to the practical management of the world's fishery management. Such models have, by revealing the negative consequences of non-cooperation, contributed to encouraging and sustaining the joint management of transboundary fishery resources in particular. Typical examples are the mutually beneficial management of the North-east Atlantic cod stock by Russia and Norway, and the joint management of the southern bluefin tuna by Australia, Japan, and New Zealand. This review has also shown that while much has been achieved through the use of game theory in analysing fishery management problems, more needs to be done. Models for the conservation and management of high sea fisheries need to be fully developed, especially with respect to determining viable cooperative outcomes. In addition, great opportunities are available for more empirical game-theoretic modeling of fisheries management problems, by combining the many solution procedures currently available in the computational and simulation literature with the ever-increasing power of computers to address important fishery management problems.

More recent developments in the application of game theory to fisheries include (i) addressing climate variability and change (e.g. Miller and Munro, 2004; Miller *et al.*, 2013); (ii) developing sequential games (e.g. McKelvey, 1997; Hannesson, 1995; Chapter 10 of this book); games with more than two players by allowing coalitions in models (Kaitala and Lindroos, 1998; Arnason *et al.*, 2000; Brasao *et al.*, 2000 and Duarte *et al.*, 2000; Lindroos *et al.*, 2007; Ishimura *et al.*, 2012, 2013) and a partition function approach, which captures the influence of group externalities (e.g. Pintassilgo, 2003). Authors are beginning to show how game theory may be used to inform group decisions in biodiversity conservation (Frank and Sarkar, 2010). Two recent complementary reviews of game theory and fisheries are Bailey *et al.* (2010) and Hannesson (2011).

3 Cooperative and non-cooperative management when capital investment is malleable[1]

Introduction

This is an applied empirical analysis of the exploitation of the North-east Atlantic cod found in the Barents Sea; the focus is on joint (cooperative) versus divided (non-cooperative) management of the stock. The North-east Atlantic cod stock (ANCS) is shared between Norway and Russia (and to a lesser extent, third countries), with approximately 45% of the present total allowable catch (TAC) to each of the two parties, and 10% to third parties. Over the past decades 45–75% of the TACs have been taken by trawlers (ICES, 1996). Russia and third countries use mainly trawl, while Norway employs mainly coastal vessels and trawl. Hence, the bulk of the ANCS is landed by coastal and trawler fishing vessels. Coastal vessels target mature cod of age groups 7 and above, while trawlers catch juveniles and mature cod of age groups 4 and above (Hannesson, 1993b). As a result of this difference, interesting game theoretic analysis can be carried out to investigate the consequences of the action of one class of vessels on (i) the economic benefit of the other; and (ii) the stock sustainability of the resource.

Our analysis assumes an organization of the Barents Sea cod fisheries around the two main fishing gears used in the exploitation of the resource. First, fishing vessels active in the fisheries are organized into two broad groups: the coastal and trawler vessel groups are considered two separate and distinct entities that can choose either to cooperate or not. Henceforth, these are denoted Coastal Fisheries Management (C) and Trawl Fisheries Management (T). The assignment of two separate and distinct fleets to the two managements captures, to some extent, the division of the stock between Norway and Russia, but even in Norway a division is usually made between the coastal fleet and the trawlers, and the Norwegian quota is divided between these. Second, within C and T, it is assumed that there are many *cooperative* agents.[2] Hence, non-cooperation can occur only at the level of C and T. The problem remains to find out what the overall annual catch for cod will be, and what proportion of this will be taken by the coastal and trawl fleets, respectively, under (i) non-cooperation; (ii) cooperation without side payments; and (iii) cooperation with side payments. In both of these cooperative regimes, it is assumed that catch shares are not predetermined.[3] Of course, predetermined catch shares could be allowed in the model; this is not done because the determination of catch shares for the two agents is one of the central objectives of this chapter.

Note that here conflicts in management strategies between C and T arise mainly through differences in fishing grounds and gear, and the age groups of cod targeted. Also, conflicts arising from unequal catching costs are important. Here, conflicts arising from perceived differences in the discount factor and consumer preferences (Munro, 1979) are ignored by assuming equal discount factors and constant price per kilogram of fish across players.

Earlier attempts to study the optimal management of joint resources have relied on static fisheries models, and have not addressed the problem of resolving conflicts of interest between the joint owners of the resource (Anderson, 1975a, b). In more recent times these issues have received the attention of some researchers, for example Levhari and Mirman (1980), Munro (1979, 1990), Armstrong and Flaaten (1991a), and Fischer and Mirman (1992). This chapter is of general interest because it develops an applied computational game theoretic model in a manner that is rare in the literature. There are two major differences between this work and Chapter 4 below. First, in contrast to the latter, capital investment is assumed to be perfectly malleable here. Second, unlike in Chapter 4, bargaining and cooperation are explicitly modeled in this study.

A number of questions are posed and explored in the present chapter: (i) What is the discounted resource rent that can be realized from the resource if (a) only C, (b) only T, and (c) both C and T exploit the resource under non-cooperation and under cooperation? (ii) What is the effect of exploitation under each of the above scenarios on the stock level? (iii) Which of the scenarios gives the optimal solution both in terms of discounted resource rent and the long-term survival of the stock? The main issue is whether the optimal solution involves C or T when they operate as sole owners, and how the optimal solution compares with the non-cooperative and cooperative game solutions with two players; (iv) How do the costs and prices faced by the players, the discount factor, their selectivity patterns, and the survival rate of the stock affect the results of the study?

A key result of the chapter is that the maximum discounted resource rent from the North-east Atlantic cod stock is achieved under cooperation with side payments. In which case C simply buys out T and exploits the resource as a sole owner. The reasons for the excellent performance of C are twofold. First, C targets only mature cod, which command a 15% price premium. Second, the almost "magical" selectivity of the coastal fleet makes for a biologically efficient targeting of cod.

In the next section, the model, a special feature of which is the explicit modeling of the biologically and economically important age groups of cod, is presented. The theoretical basis of the algorithm used to compute the equilibrium solutions predicted by the model is given in the Appendix.

The model

A two-agent bioeconomic, deterministic, dynamic game theoretic model for the exploitation of the North-east Atlantic cod is developed, which allows us to explore the economic effects of catching the resource under the different scenarios outlined above. An important assumption in the model is that capital is malleable. That is, a somewhat unrealistic assumption was made that fishers can buy and sell off fishing gear without any constraints. In a situation where fishers can (i) quickly redirect their fishing efforts to target other stocks; and/or (ii) hire their capital requirements from a rental firm for fishing vessels, then this assumption does not appear unrealistic. We relax this assumption and consider the case where capital is non-malleable in Chapter 4. Generally, the advantage of this assumption is to allow players more flexibility, and thus higher payoffs than under a non-malleable capital assumption.

Let $i = \{1, 2\}$ be the set of players in the game, where 1 denotes T and 2 denotes C; the set $\{0, \ldots, A\}$ be the age groups of fish, where A is the last age group, set equal to 15 based on the life expectancy of cod; and the set $\{1, \ldots, T\}$ denotes the fishing periods, where T is the terminal period, set equal to 32 due to computational limitations.

The demand for fish is assumed to be perfectly elastic, thus the age-dependent price per kilogram of fish, denoted by v_a, is assumed to be constant for both players. The catch cost function of a given player*p*in period t, $C(i, t)$, is modeled as an "almost" linear function of its fishing effort (number of fleets), $e_{i,t}$:

$$C(e_{i,t}) = \frac{k_i e_{i,t}^{1+\omega}}{1+\omega} \tag{3.1}$$

where $\omega = 0.01$, and $k_i/(1+\omega) \approx k_i$ is the cost of engaging one fishing fleet for one year. This formulation of the cost function has two advantages. First, it is a strictly convex cost function, which together with the linear catch function in the model gives a strictly concave objective function. This is important because strict concavity is a necessary condition for convergence of the variables in the model to their equilibrium values (Flåm, 1993). Second, by choosing a value for $\omega = 0.01$, this ensures a marginal cost of fishing effort that can be considered constant for all practical purposes.

Let the single period profit of player i be given by

$$\pi_{i,t} = \pi_i(n_t, e_{i,t}) = \sum_{a=0}^{A} v_a w_a q_{i,a} n_{a,t} e_{i,t} - C(e_{i,t}) \tag{3.2}$$

where $n_{a,t}$ is the age- and period-dependent stock size in number of fish, w_a is the weight of fish of age a, and $q_{p,a}$ is the age and player dependent catchability coefficient, that is, the share of age group a cod being caught by one unit of fishing effort. The parameter $q_{i,a}$ plays a central role in this model: it is the device used to account for the special features of our two fisheries.

Attention is focused on interactions between the players at the level of the resource. Therefore, the profit function above is formulated so as to exclude the possibility for interactions between the players in the marketplace (such interactions could, however, be easily incorporated in the model). First, a constant price means a competitive market for fish, where the quantity put on the market by any single player does not affect the price. Second, the profit function of player p is assumed to depend only on its own effort.

Non-cooperation

In this case, the problem of player p is to find a sequence of effort, $e_{p,t}$ ($t = 1, 2, \ldots, T$) to maximize the objective functional (discounted resource rent)

$$M_i(n, e_i) = \sum_{t=1}^{T} \delta_i^t \pi_i(n_t, e_{i,t}) \tag{3.3}$$

subject to the stock dynamics given by equation (3.5) below and the obvious non-negativity constraints. In the equation above, $\delta_i = (1 + r_i)^{-1}$ is the discount factor. The variable $n(n_t)$ is the post-catch stock matrix (vector) in number of fish, and r_i denotes the discount rate of player i.

Cooperation

The goal of the cooperative agents is to find a sequence of effort, $e_{i,t}$, and stock level, $n_{a,t}$, to maximize a weighted average of their objective functionals, prof_{com}.[4]

The weights β and $(1 - \beta)$ indicate how much weight is given to the own objective functionals of T and C, respectively, in the cooperative management problem. For a given $\beta \in [0, 1]$, the cooperative management objective functional translates into maximize

$$\text{prof}_{\text{com}} = \beta M_1(n, e_1) + (1 - \beta) M_2(n, e_2) \tag{3.4}$$

subject to the same constraints mentioned under non-cooperation.

The aim of this part of the analysis is to compute the cooperative discounted payoffs to the players individually and collectively for different values of β, and to determine the effort levels at which both players are likely to accept the cooperative without a side payments management solution. To do this, Nash's theory of bargaining is applied (Nash, 1953). Note that for $\beta = 1$ and $\beta = 0$, the problem reduces to that of sole ownership by T and C, respectively.

Population dynamics

Players in the game are jointly constrained by the population dynamics of the fish stock. Nature is introduced into the game with the sole purpose of ensuring that the joint constraints are enforced. The decision variable of nature is thus the stock level – its objective being to ensure the feasibility of the stock dynamics. Formally, nature's objective is expressed as 0 if the stock dynamics are feasible and $-\infty$ otherwise.

Let the stock dynamics of the biomass of fish in numbers $n_{a,t}$ be described by

$$\begin{aligned} & n_{0,t} = f(B_{t-1}) \\ & n_{a,t} + h_{a,t} \leq s_{a-1} n_{a-1,t-1}, \text{ for } 0 < a < A \\ & n_{A,t} + h_{A,t} \leq s_A n_{A,t} + \varsigma_{A-1} n_{A-1,t-1}, \text{ given } n_{a,0} \end{aligned} \tag{3.5}$$

where

$$\begin{aligned} & f(B_{t-1}) = \frac{\alpha B_{t-1}}{1 + \gamma B_{t-1}} \\ & B_{t-1} = \sum_a p_a w_{sa} n_{a,t-1} \\ & h_{a,t} = \sum_i q_{i,a} e_{i,t} n_{a,t} \end{aligned}$$

The function $f(B_{t-1})$ is the Beverton–Holt recruitment function[5]; B_{t-1} represents the spawning biomass in weight; p_a is the proportion of mature fish of age a; w_{sa} is the weight at spawning of cod of age a; α[6] and γ are constant parameters; sa is the natural survival rate of fish of age a; $h_{a,t}$ denotes the combined catch of fish of age a, in fishing season t, by both agents. In addition to the joint constraints mentioned above, players are faced with non-negativity restrictions such as $e_{i,t} \geq 0$, for all i, t; $n_{a,t} \geq 0$, for all a, t and $n_{a,T+1} \geq 0$, for all a.

Existence and uniqueness of equilibrium solutions

Typically, in games of the sort constructed here there are problems related both to the existence and uniqueness of equilibrium solutions. However, it is shown in Cavazutti and Flåm (1992) that under certain conditions open loop equilibria for the class of games under consideration do exist. In addition, the authors show that if along the equilibrium profile all players impute the same shadow prices (Lagrange multipliers) to the resource constraint, then the equilibrium tends to be unique.

It should be noted that the solutions computed in the non-cooperative scenario do not subscribe fully to the customary open loop solution concept derived from control theory. In such solutions, agents are expected to directly control the stock level too. In our non-cooperative scenario, the stock variable is controlled by nature, and only indirectly by fishers through the choice of effort level. In the cooperative and sole ownership scenarios, however, players take into direct account the marginal stock effect by maximizing with respect to the stock level.

Table 3.1 Parameter values used in the model

Age a	*Catchability coefficient q(p, a)*		*Weight at spawning w(s, a)*	*Weight in catch w(a)*[a]	*Initial numbers*[b]
(years)	*TF*	*CF*	*(kg)*	*(kg)*	*(millions)*
1	0	0	0.10	0.30	460
2	0	0	0.15	0.60	337
3	0	0	0.28	0.77	298
4	0.032	0	0.51	1.06	223
5	0.062	0	0.99	1.55	117
6	0.075	0	1.72	2.27	61
7	0.084	0.056	2.86	3.57	33
8	0.084	0.140	4.68	5.12	9
9	0.084	0.191	6.61	6.61	9
10	0.084	0.255	7.29	7.29	9
11	0.084	0.217	8.91	8.91	9
12	0.084	0.153	10.85	10.85	9
13	0.084	0.089	12.50	12.50	9
14	0.084	0.051	13.90	13.90	9
15	0.084	0.0255	15.00	15.00	9

The parameter p_a is given the value (0, 0, 0, 0, 0.02, 0.06, 0.25, 0.61, 0.81, 0.93, 0.98, 1, 1, 1, 1) for $a = \{0, 1, \ldots, 15\}$.

a Both $w(a)$ and $w(s, a)$ are taken from ICES (1996).

b These are obtained by taking average initial numbers of various age groups from 1984 to 1991 reported in Table 3.12 of the ICES Report 1992).

Data

Table 3.1 lists the parameter values used for the computations.[7] In addition, α and γ are set equal to 1.5 and 1 per billion kilograms, respectively, to give a billion zero age fish when the spawning biomass is two million tonnes.[8] Based on the survival rate of cod, s_a is given a value of 0.81 for all a. The price parameter, v_a, is set equal to Norwegian kroner (NOK)[9] 6.78 for age groups 0 to 6 and NOK 7.46 for age groups greater than 15 years.[10] The cost parameter, k_i, which denotes the cost of engaging a fleet of vessels (10 and 150 for T and C, respectively) for one year, is calculated to be NOK 210 and 230 million for T and C, respectively.[11] The discount rate, r_i, is set equal to 7% for all i, as recommended by the Ministry of Finance of Norway. The initial numbers of cod of age groups 1 to 8 are obtained by taking the average of the initial numbers from 1984 to 1991, reported in Table 3.12 of the ICES (1992). For the other age groups, the same number as for age group 8 is assumed. This gives an estimated initial stock size of 2.24 million tonnes. These numbers are the average percentage age at maturity over 1990–1995, reported in Table 3.8 of ICES (1996).

Table 3.2 Discounted profit to the agents for different scenarios (in billion NOK)

	β						
	1.0	*0.8*	*0.7*	*0.6*	*0.25*	*0.0*	*Non-cooperative*
TF	56.14	33.65	32.82	30.39	23.80	0	27.42
CF	0	18.82	21.49	21.77	26.19	58.78	19.70
Total	56.14	52.46	54.31	52.17	49.99	58.78	47.12

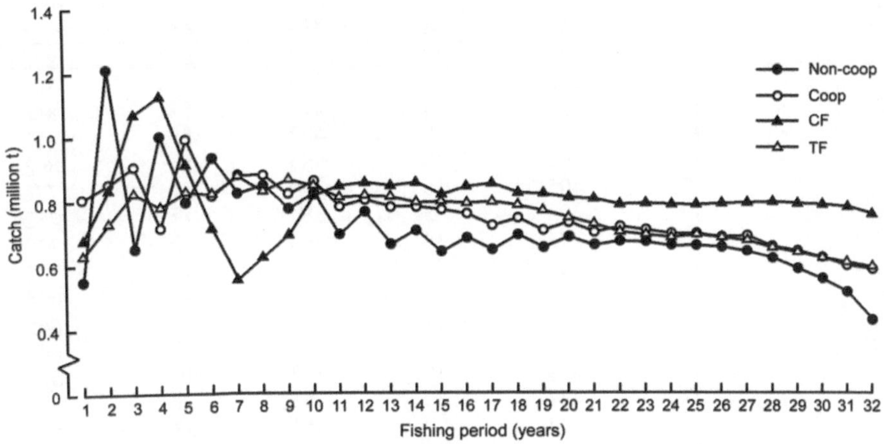

Figure 3.1 Catch profiles for the different scenarios.

Results

Results of the computations are given in Tables 3.2 and 3.3 and Figures 3.1 and 3.2. To obtain these, dynamic simulation software package, Powersim, is used as computational support.[12]

Payoffs

Table 3.2 gives both the non-cooperative and cooperative (including sole ownership) equilibrium solutions. In the case of the latter, outcomes for $\beta = 1$, 0.8, 0.7, 0.6, 0.25 and 0 are given. It is clear from this table that the best economic result is obtained when β is equal to 0. That is, when the preferences of C are given full consideration at the expense of those of T. Also, the table shows that non-cooperation produces the worst economic result. The scenarios represented by $\beta = 0$ or 1 can be realized only if agents agree to cooperate with side payments, in which case it will be economically optimal for C to buy out T in order to

Table 3.3 Number of vessels employed by the agents under different scenarios

	Non-cooperative		*Cooperative*		*Sole ownership*	
Period (t)	*TF*	*TC*	*TF*	*TC*	*TF*	*TC*
1	54	752	49	480	61	963
2	58	812	54	537	70	1409
3	57	779	53	530	72	1382
4	56	737	52	518	72	1326
5	54	693	50	495	72	1250
6	52	656	49	475	71	1172
7	50	630	47	465	69	1098
8	48	609	45	449	67	1034
9	47	588	44	437	64	981
10	45	572	42	429	62	938
11	43	555	41	420	59	899
12	42	539	39	410	57	866
13	41	524	38	402	55	833
14	39	509	37	395	53	803
15	38	495	36	390	51	776
16	37	485	35	384	49	752
17	36	474	34	380	47	728
18	35	467	33	374	46	705
19	34	459	32	369	44	684
20	34	453	31	365	43	665
21	33	444	31	360	41	647
22	32	438	30	357	40	629
23	31	431	30	354	39	611
24	31	423	29	351	38	594
25	30	417	28	347	37	579
26	30	411	28	344	36	564
27	29	404	27	342	35	551
28	29	398	27	339	34	537
29	28	390	27	336	33	524
30	28	384	26	333	32	510
31	27	375	26	330	31	497
32	26	359	25	327	30	482

operate the fishery under sole ownership. T is then compensated by receiving its "threat point," which here is the Nash non-cooperative outcome, plus 50% of the surplus over the total non-cooperative discounted rent to both players (Nash, 1953).

In the more realistic case where society may rather keep the two fisheries in operation, the more interesting question to ask is, what β are the agents likely to agree upon in a cooperative without a side payment arrangement in which catch shares are not predetermined? Applying Nash's theory of bargaining (Munro, 1979; Binmore, 1982; Kaitala, 1986), the players will settle for $\beta = 0.7$,

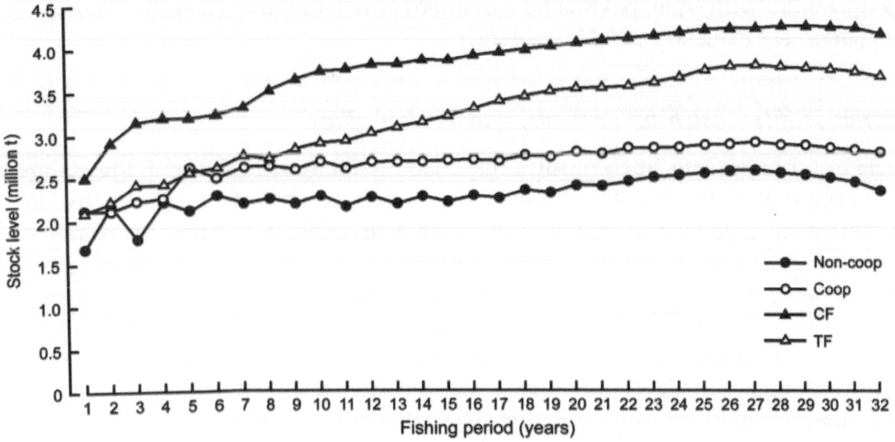

Figure 3.2 Stock profiles for the different scenarios.

as this maximizes:

$$(PV_1^{\text{Coop}} - PV_1^{\text{NC}})(PV_2^{\text{Coop}} - PV_2^{\text{NC}}) \tag{3.6}$$

where PV denotes present value rent, the subscripts 1, and 2 refer to T and C, respectively, and the superscripts Coop and NC stand for cooperative and non-cooperative, respectively.

An interesting observation that emerges from Table 3.2 is that under sole ownership C produces the best results, but once both players catch the stock under non-cooperation, T makes the higher rent. This means T has more bargaining power than C in the competitive situation. As a result, T comes out better in the cooperative without side payments arrangement. This result captures some of the dilemmas that can be faced by managers of joint resources: even though one agent may be the best in terms of, say, the price it can achieve for its catch, or the way it targets the resource, it may happen that the other party has the higher bargaining power in a game situation.

In more concrete terms, when both players exploit the resource under non-cooperation, the total discounted rent accruable is NOK 47.12 billion, that is, the sum of the discounted rents of the two players. Of this amount, T makes NOK 27.42 billion, while C earns NOK 19.70 billion. In comparison, a cooperative without side payments arrangement results in total discounted rent of NOK 54.31, with T netting NOK 32.82 and C making NOK 21.49. Table 3.2 also reveals that under sole ownership, T obtains NOK 56.14 billion, while C makes NOK 58.78 billion. Hence, under cooperation with side payments and no predetermined catch shares, T and C will make NOK 33.25 and NOK 25.53, respectively. Note that by moving from non-cooperation to cooperation with side payments, an improvement in the overall discounted resource rent of nearly 25% is achieved.

Effort profiles, catch proportions, and stock profiles

The effort levels and stock profiles that underlie the results above are presented in Tables 3.3 and 3.4 and Figures 3.1 and 3.2. The effort levels for the different management scenarios are reported in Table 3.3. The superiority of the cooperative management strategies seems to stem partly from the reduction of vessel capacity when management is changed from divided to unified management (Table 3.3). Average catch per unit effort (CPUE) for the trawlers and coastal vessels are 9,930 and 592 tonnes under non-cooperation; 12,017 and 769 tonnes under cooperation without side payments; and 14,731 and 997 tonnes under sole ownership.[13] The relatively low CPUE numbers under non-cooperation are a clear indication of the inefficiency inherent in that case.

An examination of the catches of T and C suggests that both under non-cooperative and cooperative management, T catches a larger proportion of the total catch over the years, specifically, the model tends to support giving T an average of about 60% of the total annual catch. This reflects the strategic advantage that T has over C.

Figure 3.1 gives an idea of the absolute total catches in each year. It is seen from the figure that catch levels tend to be unsettled in the early period of the game, but stabilize somewhat by the 12th fishing period. C, as sole owner, produces the highest catch levels, while the non-cooperative regime produces the lowest catches.

Figure 3.2 illustrates graphically the stock profiles that emerge from the use of the effort levels given in Table 3.3. The middle parts of the graphs give an indication of long-run behavior of biomass under the different management scenarios. As expected, the stock profiles are higher under cooperation and sole ownership, with the highest profile obtained when C has sole ownership.

Intuitively, these results can be explained: both players, knowing that if they let fish escape now, they will be the only ones to catch it tomorrow, have a better incentive to do so when they have sole fishing rights over the resource. The positive effects of better conservation, or the gains due to the elimination of the "tragedy of the commons" is expected to have a positive effect on the discounted rent accruable to both players, which by extension leads to higher discounted rent to the (fishing) community as a whole.

It should be noted that contrary to expectations, the fish population is not driven to extinction at the end of the game. There are a number of possible reasons for this. First, players do not target all age groups. Second, economic extinction is not necessarily the same as biological extinction. Third, it appears the 32-year time horizon is enough to let players behave as if they were facing an infinite time horizon. Indeed, sensitivity analysis using a time horizon of 15 years tends to show that players exert more pressure on the stock at the end of the game

Table 3.4 Effect of key parameters on overall discounted rent from the resources (in billion NOK)

Management alternative	*Base case*	*25% ↑[a] in price*	*25% ↓ in cost*	*↓ in DF (0.935–0.91)*	*Same price[b]*	*Knife-edge selectivity[c]*	*↑ in adults (initial stock)[d]*	*↑ in juveniles (initial stock)[e]*	*↑ in SR[f] (0.81 to 0.9)*	*Time horizon = 15 years*
Non-cooperative										
TF	27.42	42.92	31.90	20.14	25.22	22.70	32.95	32.84	45.68	19.66
CF	19.70	30.66	24.66	14.75	16.85	29.11	24.51	17.93	26.49	13.05
Total	47.12	73.58	56.56	34.89	42.07	51.82	57.46	50.77	72.17	32.17
Relative profitability[g]	**0.72**	**0.71**	**0.77**	**0.73**	**0.67**	**1.28**	**0.74**	**0.55**	**0.58**	**0.66**
Sole owner										
CF	58.78	99.15	63.82	36.46	47.54	70.43	63.86	66.93	117.12	38.93
TF	56.14	81.50	61.50	37.84	50.32	51.89	66.20	65.25	91.53	39.46
Relative profitability	**1.05**	**1.22**	**1.04**	**0.96**	**0.94**	**1.36**	**0.96**	**1.23**	**1.28**	**0.99**

[a] The arrows ↑ and ↓ mean "increase" and "decrease," respectively.
[b] Price is NOK 6.78 for all age groups.
[c] Set equal to 1 for age groups >3, in the case of trawlers, and age groups > 6, in the case of coastal vessels, otherwise set to 0.
[d] Five times more mature cod, this is little to start with.
[e] Two times more juveniles.
[f] SR, survival rate.
[g] Relative profitability defined as discounted rent to CF divided by that to TF.

when the time horizon is short. The first two points apply more to C, hence, the high conservation of the stock when only C exploits.

Sensitivity analysis

Table 3.4 gives the discounted rent to C and T given changes in key parameter values. Note that column 2 of the table gives the "base case" outcome, while the rest of the columns give the outcomes for given *ceteris paribus* changes in various parameters of the model. The rows to look at closely are those for relative profitability, which is defined as the profit to C divided by that to T.

The following key observations can be made from Table 3.4:

Under non-cooperation,
(i) T does better than C in all cases except when knife-edge selectivity is assumed for both vessel types: note that this assumption takes away most of the advantage that T has over C (in the competitive situation) due to its selectivity pattern. (ii) In comparison to the "base case" scenario, T does relatively better than C with a decrease in time horizon, increase in the proportion of juveniles in the initial stock, increase in the survival rate of cod, when the price premium on mature cod is taken away, and when there is an across-the-board increase in price. On the other hand, C does relatively better than T with increasing discount factor, decrease in the unit cost of renting vessels, and increase in the proportion of mature cod in the initial stock.

Under sole ownership,
(i) C does better than T in all cases except (a) when there is an across the board decrease in discount factor, (b) when the price premium for mature cod is taken away, (c) when there is an increase in the proportion of mature cod in the initial stock, and (d) when there is a decrease in the time horizon of the game; (ii) in comparison to the "base case" scenario, C does relatively better than T with an increase in price, when knife-selectivity is assumed, with an increase in the survival rate and the proportion of juveniles in the initial stock. On the other hand, a decrease in the unit cost of renting vessels tends to favor the trawler fleet.

Concluding remarks

Under the assumptions of our model and available data, maximum discounted rent from the North-east Atlantic cod stock is achieved under a cooperative with side payments arrangement, where catch shares are not predetermined. In which case C simply buys out T and manages the resource solely. However, sensitivity analysis shows that T can do the buying out if (i) the price per unit weight of cod is assumed to be age-independent, (ii) agents are impatient (that is, with a decrease in the discount factor), (iii) players have a short time horizon, and (iv) there is an increase in the proportion of mature cod in the initial stock.

A point that needs to be made here is that the high conservation of the stock that occurs when C exploits the resource as sole owner would almost surely have worked in favor of the coastal fleet had our study been stochastic. Recall that several stochastic bioeconomic studies call for what is termed the *precautionary approach* to resource management, where caution is called for in the choice of catch rates in order to build up stocks to levels that are in a better position to cope with uncertainty (Andersen and Sutinen, 1984).

The results of the current chapter concords with the findings of earlier studies of the ANCS, e.g. Hannesson (1978) and Armstrong *et al.* (1991) tend to favor the coastal fleet. However, the superior performance of C in the present study hinges mainly on the fact that mature cod commands a price premium. The results of Armstrong and Flaaten (1991b), which studies the ANCS in the context of the former Soviet Union and Norway, that cooperation is the economically better management regime, concords with the findings of this chapter, and with theoretical results on the management of shared resources (Munro, 1979).

Possible extensions of the work in this chapter include (i) relaxing the malleable capital assumption in the model, which is done in Chapter 4 below; (ii) introducing some form of interaction at the market place[14]; (iii) introducing uncertainties; and (iv) undertaking multispecies analysis to capture the natural interaction between cod and capelin in the Barents Sea, as done in Chapter 5.

4 Cooperative and non-cooperative management when capital investment is non-malleable[1]

Introduction

This chapter considers non-cooperative use of a common property fish stock, namely, the North-east Atlantic cod. Attention is focused on a restricted access fishery where only two agents participate in the exploitation of the resource, the aim being to predict the number of vessels that each agent in such a situation will find in his or her best interest to employ. An important although self-evident aspect of the game is that both agents are jointly constrained by the population dynamics of the resource. The key assumption of the chapter is that players undertake investment in capital that is irreversible. This assumption is quite realistic because capital embodied in fishing vessels is often non-malleable: non-malleability is used here to refer to the existence of constraints upon the disinvestment of capital assets utilized in the exploitation of the resource (Clark *et al.*, 1979). This implies that once a fishing firm or authority invests in a fleet of vessels it either has to keep it until the fleet is depreciated, or else the vessels can only be disposed of at considerable economic loss.

A number of papers have appeared in the fishery economics literature that focus, among other things, on the irreversibility of capital employed in the exploitation of fishery resources. Examples include Clark *et al.* (1979), Clark and Kirkwood (1979), Dudley and Waugh (1980), Charles (1983a, 1983b), Charles and Munro (1985) and Bjørndal and Munro (2012). I am, however, not aware of any prior work that models, computes numerically, and analyses the exploitation of fishery resources as done in this chapter. Among the examples cited above, only Dudley and Waugh (1980) consider investment decision in a fishery with more than a single agent participating. But even in this case, only qualitative statements of the likely effects of this are made. The study by Clark and Kirkwood (1979) is close to the work planned herein, at least in terms of the kinds of questions they address. The authors presented a bioeconomic model that predicts the number of vessels of each of the two types entering the prawn fishery of the Gulf of Carpentaria under free access. In addition, they estimated the economically optimal number of vessels of each type. The results they obtained are then compared with available data on the prawn fishery of the Gulf of Carpentaria.

These are also issues addressed in this contribution, albeit with a number of differences. First, there is a difference with respect to the number of agents in the two studies: While Clark and Kirkwood (1979) consider the social planner's and open access equilibrium fleet sizes, equilibrium fleet sizes that will emerge in a non-cooperative environment involving two agents are computed, and then, using these results, the social planner's equilibrium fleet size is derived and the probable open access equilibrium fishing capacity is discussed. Thus, this contribution adds a new dimension to the discussion, namely, the two-agent analysis. Second, there is a difference in the way the population dynamics of the fish stock is modeled: while their study prescribes and uses a single cohort to describe the fish stock, a multi-cohort population structure is accommodated.

The detailed concern of this study is to develop the necessary framework to:

1. identify a Nash non-cooperative equilibrium solution for a bimatrix game involving the trawl and coastal fisheries operating on the North-east Atlantic cod;
2. identify the sole owner equilibrium solutions for the two fisheries, and determine which among these gives the optimal solution;
3. compare the results in (1) and (2) above to (i) the results in Chapter 3, where perfect malleability of capital is assumed implicitly, and (ii) with available data on the North-east Atlantic cod. The former comparison would put us in a position to say something about the possible gains of establishing rental firms for fishing vessels and/or allowing mobility of vessels between different stocks;
4. discuss the fishing capacities that are likely to emerge in an open access scenario; and
5. investigate the effect of fixed cost, discount rates, initial stock size, and the terminal constraint, on the relative profitability of the players.

The next section gives a brief description of the North-east Atlantic cod fishery. The third section presents the model, a special feature of which is the explicit modeling of the biologically and economically important age groups of cod. This is followed by a brief mention of the algorithm for the computation of the equilibrium solutions, the theoretical basis of which is given in the Appendix. Next, the results of the study are stated, followed by the conclusions..

The North-east Atlantic cod fishery

The North-east Atlantic cod is a stock of Atlantic cod, arguably among the world's most important fish species. It inhabits the continental shelf from shoreline to 600 m depth, or even deeper, usually 150 to 200 m. It is gregarious in behavior, forming shoals or schools and undertaking spawning and feeding migrations. The diet of adult cod is variable and consists mainly of herring, capelin, haddock, and codling. The North-east Atlantic cod spawns only along the Norwegian coast, mainly in Lofoten in April–March. Typically, it starts

Table 4.1 Number of Norwegian vessels operating on the cod fishes group for five different years

Year	*Trawlers*	*Coastal vessels*
1991	57	562
1990	51	572
1988	84	661
1986	118	628
1984	128	718

spawning at the age of 7–8 years; eggs are carried by the Gulf Stream, over the coast where they hatch, and into the Barents Sea up towards Svalbard where the young cod grow. It has a relatively long life span: it can live for well over 15 years. A majority of young cod die quite early, either because of a lack of adequate food, or because they are eaten up by other fishes. Young cod between the ages of 3–6 come to the Finnmark's coast every year. This is because mature capelin, which cod preys (the subject of Chapter 5) on, move to their spawning spots close to the Finnmark's coast. Cod follows and feeds on them, thus resulting in good spring cod in the period April to June.

The North-east Atlantic cod stock is a shared resource, jointly managed by Norway and Russia. Norwegian fishers employ mainly coastal and trawl fishery vessels in the exploitation of the resource, while their Russian counterparts employ mainly trawlers. Table 4.1 gives the number of Norwegian trawl and coastal fishery vessels (of 13 m length and over) that operated on the "cod fishes group"[2] for five different years. In addition to this comes the part of the fishing capacity employed to exploit other species, say, the "herring fishes group," which also lands the cod fishes as bycatch.

Using Norwegian data,[3] the number of coastal fishery vessels and trawlers used by Norwegian fishers in the exploitation of the cod fishes group in 1991 was calculated to be about 638 and 58, respectively. These landed about 130 and 270 thousand tonnes of cod, respectively.

To facilitate our analysis, three simplifications (about this fishery) are made.[4] First, only Norwegian prices and costs are used in the analysis. Second, the vessel types employed in the exploitation of the resource are grouped into two broad categories, namely, the coastal and the trawl fisheries, and placed under the management of two separate and distinct management authorities, henceforth to be known as Coastal Fisheries Management (C), and Trawl Fisheries Management (T). Third, only the *most cost effective* vessels[5] in each of these categories are assumed to be employed in the exploitation of the resource. The assignment of two separate and distinct fleets to the two management authorities captures, to some extent, the division of the stock between Norway and Russia, but even in Norway a division is usually made between the coastal fleet and the trawlers, and the Norwegian quota is divided between these.

The model

The model presented here builds on that discussed in Chapter 3 to which the reader is referred for details. Here, a two-stage, two-player, dynamic, deterministic, non-cooperative game model is put together, the two players being T and C. By a game we mean a normal (strategic) form, simultaneous-move game in which both players make their investment decisions in ignorance of the decision of the other. At stage one of the game, each player invests in fishing capacity *ex ante*, bearing in mind that such investment is irreversible. Then in stage two, the players employ their chosen capacity investment to exploit the shared resource for the next 15 years, subject to the stock dynamics and non-negativity constraints.

Both T and C are assumed to be rational and act here to maximize their discounted profit (payoff) function $\Pi_i : K_T \times K_C \rightarrow \Re$ where K_T and K_C are the pure strategy sets of player $i = \mathrm{T}, \mathrm{C}$, that is, the set of fishing capacity (number of vessels or fleet size) that a player can choose from. Player i's payoff at an outcome (k_T, k_C) is then given by $\Pi(k_T, k_C)$. A major aim of this modeling exercise is to find the strategy pair (k_T^*, k_C^*) such that no player will find it in his or her interest to change strategy given that his or her opponent keeps to his. In other words, the goal is to find a *Nash non-cooperative equilibrium* in a two-player fishery game, where k_T^* is a best reply to k_C^* and vice versa. This is equivalent to stipulating that the inequalities

$$\begin{aligned} \Pi_\mathrm{T}(k_\mathrm{T}^*, k_\mathrm{C}^*) &\geq \Pi_\mathrm{T}(k_\mathrm{T}, k_\mathrm{C}^*) \\ \Pi_\mathrm{C}(k_\mathrm{T}^*, k_\mathrm{C}^*) &\geq \Pi_\mathrm{C}(k_\mathrm{T}^*, k_\mathrm{C}) \end{aligned} \tag{4.1}$$

hold for all feasible k_T, and k_C.

On existence of the Nash equilibrium

Nash (1950b, 1951) proved the existence of equilibrium points under certain assumptions on each player's strategy space and corresponding payoff function. Essentially, he dealt with matrix games. Rosen (1965) went further to show that when every joint strategy lies in a *convex, closed,* and *bounded* region in the product space and each player's payoff function Π_i, $i = \mathrm{T}, \mathrm{C}$ is *concave* in his or her own strategy and *continuous* in all variables, then there is at least one Nash equilibrium of the game. This result is stated in theorem 1 below.

THEOREM 1 (Existence of Nash equilibrium. Rosen (1965)): *An equilibrium point exists for every concave* n*-person game.*

The game formulated here is a concave two-person game, and hence satisfies the above theorem. Hence, at least one Nash equilibrium is expected to exist.

On uniqueness of Nash equilibrium

Two steps are taken here to deal with the vexing problem of equilibrium selection.

Step 1: Only open loop strategies are allowed in the second stage of the game.[6] That is, each player commits, in advance, his or her fishing capacity to a fixed time function rather than a fixed control law (closed loop strategies). Note that unlike in the case of fixed control laws, where the choice of control depends on the past history of the game, fixed time functions are independent of the actions of the opponent so far in the game. In the information theoretic sense, open loop corresponds to the receipt of no information during play, while closed loop represents full information.

The main reason open loop strategies are computed, even though they are not likely to lead to close form solutions, is that it would be practically impossible to compute the predictions of our model if closed loop strategies were allowed. This is because closed loop strategies normally entail complex and huge numbers of strategies in repeated games (Binmore, 1982). Another reason is that the new "Folk Theorem for Dynamic Games" introduced by Gaitsgory and Nitzan (1994), gives us reason to believe that under certain monotonicity assumptions, the set of closed loop solutions that may emerge from our model may coincide with the open loop solutions computed herein.

Step 2: The same shadow prices (that is, Lagrange multipliers) are imposed across both players for resource constraint violation. Flåm (1993) shows that in addition to this, if the marginal profit correspondence is *strictly monotonic*, then there exists a unique Nash equilibrium for our game. Incidentally, strict monotonicity of marginal profit correspondence is also a sufficient condition for convergence in our model.

In the presentation of the mathematical equations in the rest of the model, two other subscripts ($a = 0, \ldots, A$, and $t = 1, \ldots, T$) are used to denote age groups of fish, and time periods or stages, respectively.[7] Based on the life expectancy of cod, the last age group A, is set equal to 15. The finite time horizon of the game, T, is set equal to 15 due to computational limitations.

Catch

Let catch of age group a (in number of fish) by player i in fishing period t, $h_{i,a,t}$, be given by

$$h_{i,a,t} = q_{i,a} n_{a,t} E_{i,t} \tag{4.2}$$

where the effort profile, $E_{i,t} = k_i e_{i,t}$, and k_i is the *ex ante* fixed capacity investment of player i; $e_{i,t} \in [0, 1]$, is the capacity utilization, that is, the fraction of k_i taken out for fishing in a given year; $n_{a,t}$ is the post catch number of fish of age a in fishing period t and $q_{i,a}$ is the player and age-dependent catchability coefficient, that is, the share of age group a cod being caught by one unit of effort.

Total catch by all players of age group a in period t can thus be written as

$$h_{a,t} = \sum_i q_{i,a} n_{a,t} k_i e_{i,t} \tag{4.3}$$

Total catch in weight by all players over all age groups in period t is given by

$$h_t = \sum_a h_{a,t} w_a \tag{4.4}$$

where w_a is the weight of fish of age group a.

Costs and prices

The fishery is assumed to face perfectly elastic demand. Thus, the ex-vessel selling price of fish per kilogram, v, is assumed to be constant. The fishing costs per vessel employed by player i in period t, $\psi_{i,t}$, consist of fixed costs (ϕ_i) and variable costs (ξ_i) which are proportionate to $e_{i,t}$

$$\psi_{i,t} = \varphi_i + \frac{\xi_i}{(1+\omega)} e_{i,t}^{1+\omega} \tag{4.5}$$

where $\omega = 0.01$. This formulation of the cost function ensures strict concavity of individual profit as a function of individual effort. This strictness is important for the sake of convergence.

Revenue

The revenue to player i, in period $t, r_{i,t}$, comes from the sale of catch over all age groups in that period, that is,

$$r_{i,t} = v \sum_a w_a h_{i,a,t} \tag{4.6}$$

Profits

Player T's profit in a given period $s$is then given by the equation

$$\pi_{i,t}(n_t, e_{i,s}) = r_{i,t} - k_i \psi_{i,t} \tag{4.7}$$

where $k = (k_i, k_{-i})$. Note that $\pi_{i,t}$ is a function of the actual fish abundance in a period, n_t, and own effort in that period. The analysis was restricted to the case of perfectly non-malleable capital in which the depreciation rate is equal to zero and capital has a negligible scrap value. Even though this simplification is not quite realistic, the qualitative effect of this is expected to be insignificant. The profit function given by equation (3.7) is formulated to incorporate this restriction.

Objective

Given k_i, the second stage problem of player i is to find a sequence of capacity utilization, $e_{i,t}$ ($s = 1, 2, \ldots, T$), to maximize his or her present value (PV) of profits (payoff), that is,

$$\Pi_i(k_T, k_C) = \max_{e_i} \sum_{t=1}^{T} \delta_i^t \pi_i(n_t, e_{i,t}) \tag{4.8}$$

subject to the stock dynamics and the obvious non-negativity constraints, expressed mathematically as $e_{i,t} \geq 0$, for all i and t; $n_{a,t} \geq 0$, for all a and t; $n_a, T+1 \geq 0$, for all a; and $n_a, 0 \geq 0$ given. Here, $\delta_i = (1+r_i)^{-1}$ is the discount factor; $r_i > 0$ denotes the discount rate of player i; and $n_{a,0}$ is a vector representing the initial number of fish of each age group.

An important but self-evident component of this game is that players are jointly constrained by the population dynamics of the fish stock. Nature is introduced (as a player) in the game with the sole purpose of ensuring that the joint constraints are enforced. The decision variable of nature is thus the stock levels – its objective being to ensure the feasibility of the stock dynamics. Formally, nature's objective is expressed as 0 if the stock dynamics are feasible and $-\infty$ otherwise.

It is worth mentioning here that unless players enjoy bequest, they will typically drive the fishable age groups of the stock to the open access equilibrium level at the end of the game, if the terminal restriction is simply $n_{a,t} \geq 0$, for all a. To counteract this tendency, one can exogenously impose the more restrictive constraint, $n_{a,T+1} \geq \check{n}_a$ where $\check{n}_a$ is a certain minimum level of the stock of age group a that must be in the habitat at the end of period $T+1$. This is what is done here. Alternatively, this restriction can be imposed endogenously by obliging the players to enter into a stationary regime maintaining constant catches and keeping escapement fixed from T onwards.

Stock dynamics

Let the stock dynamics of the biomass of fish in numbers $n_{a,t}$ (that is, the joint constraint mentioned above) be described by

$$\begin{aligned}
&n_{0,t} \leq f(B_{t-1}) \\
&n_{a,t} + h_{a,t} \leq s_{a-1} n_{a-1,t-1}, \quad \text{for } 0 < a < A \\
&n_{A,t} + h_{A,t} \leq s_A n_{A,t-1} + s_{A-1} n_{A-1,t-1}, \quad \text{given } n_{a,0}
\end{aligned} \tag{4.9}$$

where $f(B_{t-1}) = -1 \,/(1+\gamma B_{t-1})$ is the Beverton–Holt recruitment[8] function; $B_{t-1} = \Sigma_a p_a w_{at} n_{a,t-1}$ represents the post-catch biomass in numbers; p_a is the proportion of mature fish of age a; w_{at} is the weight at spawning of fish of age a; α[9] and γ[10] are constant parameters chosen to give a maximum stock size of about 6 million tonnes – a number considered to be the carrying capacity of the habitat; s_a is the natural survival rate of fish of age a; and $h_{a,t}$, defined earlier, denotes the total catch of fish of age a, in fishing season t by all agents.

Numerical method

An algorithm is developed to resolve the second stage game problem, that is to find the answer to the following question: given the fixed capacity choice of the players in stage one of the game, what level of capacity utilization should they choose in each fishing period so as to maximize their respective economic benefits? For detailed discussions of the theoretical basis for the algorithm, see Flåm (1993) and the Appendix.

Suppose, for illustrative purposes, that all constraints (except non-negativity ones) are incorporated into one concave restriction of the form $\Phi(n,e_i,e_{-i}) \geq 0$, where e_{-i} is the profile of capacity utilization of i's rival, e_i is the equivalent for player i, and n is the stock profile (note that $n \in R^{(A+1),t}$, and $e_i \in R^t$ are large vectors, hence, the a and t subscripts are ignored here). It can be stated the payoff function of player i is as follows

$$\Pi_i(k) = \Pi_i(k_i k_{-i}) = \max_{e_i} L_i(n^*, e_i, e^*_{-i}, y^*, k) \tag{4.10}$$

where

$$L_i(n, e_i, e_{-i}, y, k) = \sum_{t=1}^{T} \delta_i^t (r_{i,t} - k_i \psi_{i,s}) + y\Phi^-(n, e_i, e_{-i}, k) \tag{4.11}$$

is a modified Lagrangian, y is the Lagrange multiplier, (e_i^*, n^*, y^*) are equilibrium solutions of the variables in question, and Φ^- is given by $\min(0,\Phi)$. The adjustment rules in the algorithm are then given by

$$e_i = \frac{\partial L_i(n, e_i, e_{-i}, y, k)}{\partial e_i}, \forall i \tag{4.12}$$

$$y = \frac{\partial L_i(n, e_i, e_{-i}, y, k)}{\partial y} = \Phi^- \tag{4.13}$$

$$n = \frac{\partial L_i(n, e_i, e_{-i}, y, k)}{\partial n} = y\frac{\partial \Phi^-}{\partial n} \tag{4.14}$$

where $\partial L_i(.)/\partial e_i$ and $\partial L_i(.)/\partial y$ are the partial derivatives of $L(.)$ with respect to e_i and y, and $\partial\Phi^-/n$ is the partial derivative of the constraint function with respect to n.

The algorithm then comes in differential form: starting at arbitrary initial points (e_i, y, n), the dynamics represented by the adjustment rules are pursued all the way to the stationary points (e_i^*, y^*, n^*). Such points satisfy, by definition, the steady-state generalized equation system:

$$0 = \Phi^{-}(n, e_i, e_{-1}, k) \tag{4.15}$$

$$0 \in \partial e_i \left[\sum_{1}^{S} \delta_i^S (r_{i,s} - k_i \psi_{i,s}) + y\Phi^{-}(n, e_i, e_{-1}, k) \right], \forall i \tag{4.16}$$

with $y^* \geq 0$. It is standard form mathematical programming that individual optimality and stock feasibility are then satisfied. The numerical scheme uses Euler's method to integrate (4.12), (4.13), and (4.14) all the way to equilibrium solutions (e_i^*, y^*, n^*).

Numerical results

A strategic form for our game is given in Table 4.5. To obtain these results, the newly developed dynamic simulation software package, Powersim is used as computational support. The parameter values listed in Table 4.2 are used for the computations. In addition, α and γ are set equal to 1.01 and 1.5, respectively, to give a maximum biomass of 6 million tonnes for a pristine stock. Based on the survival rate of cod, ς_a is given a value of 0.81. The price parameter, v, is set equal to NOK 6.78. The variable costs (φ_T & φ_C) and fixed costs (ϕ_T & ϕ_C)

Table 4.2 Values of parameters used in the model

Age a	*Selectivity q(p, a)*		*Weight at spawning w(s, a)*	*Weight in catch w(a)*	*Initial numbers*
(years)	*P=1*	*P=2*	*(kg)*	*(kg)*	*(millions)*
0	0	0	0.090	0.10	167.0
1	0	0	0.270	0.30	135.0
2	0	0	0.540	0.60	108.0
3	0	0	0.900	1.00	88.3
4	0.0074	0	1.260	1.40	71.7
5	0.0074	0	1.647	1.83	58.3
6	0.0074	0	2.034	2.26	46.7
7	0.0074	0.00593	2.943	3.27	38.3
8	0.0074	0.00593	3.843	4.27	30.8
9	0.0074	0.00593	5.202	5.78	25.0
10	0.0074	0.00593	7.164	7.96	20.3
11	0.0074	0.00593	8.811	9.79	16.7
12	0.0074	0.00593	10.377	11.53	13.3
13	0.0074	0.00593	12.456	13.84	10.8
14	0.0074	0.00593	13.716	15.24	8.7
15	0.0074	0.00593	14.706	16.34	7.0

Notes

(1) The values for $q(p, a)$ are calculated using the procedure outlined in Sumaila (1994).

(2) Player T exploits fish of age 4 and above and player C fish of age group 7 and above (Hannesson, 1993).

for engaging a vessel are calculated to be (NOK 12.88 & 0.88) and (NOK 15.12 & 0.65) million for T and C, respectively.[11] The discount rate, r_i, is set equal to 7% as recommended by the Ministry of Finance of Norway. The initial number of cod of each age group is calibrated using the 1992 estimate of the stock size of cod in tonnes.[12]

In Table 4.3, rows represent player T's pure strategies k_T and columns represent player C's pure strategies k_C. Player T's payoff is placed in the southwest corner of the cell in a given row and column and the payoff to player C is placed in the north-east corner. The best payoff for player T in each column and the best payoff to player C in each row are boldfaced. As an example, notice that 19.4 has been boldfaced in cell $k_T = 65$ and $k_C = 500$. since 19.4 lies in row $k_T = 65$, this tells us that pure strategy $k_T = 65$ is player T's best reply to a choice of pure strategy $k_C = 500$ by player C.

Notice that the only cell in Table 4.3 that has both payoffs boldfaced is that which lies in row $k_T = 57$ and column $k_C = 1050$. Thus, the only pure strategy pair that constitutes a Nash equilibrium is (57, 1050). Each strategy in this pair is a best reply to the other. This gives a total PV of economic benefit equal to $\Sigma_i \Pi_i$ (57,1050) = NOK 25.87 billion, with Π_T (57,1050) = NOK 11.84 billion and Π_C (57,1050) = NOK 14.03 billion, respectively.

The overall PV of resource rents from the fishery as a function of k_T and k_C are given in Table 4.4. The entries in each cell of this table are simply the sum of the entries in corresponding cells in Table 4.3. These results indicate an optimal fleet consisting of 1100 coastal vessels and no trawlers with a PV of resource rent equal to NOK 36.11 billion, or NOK 32.83 million per vessel. In contrast, a discounted profit-maximizing fleet, consisting solely of trawlers, would be made up of 70 vessels and earn a PV of resource rent of NOK 32.42 billion, or NOK 463.14 million per vessel.

The model thus appears to support the general theoretical assertion that non-cooperation generally results in rent dissipation through the use of excess fishing capacity. The economic theory of fisheries predicts that in an open access fishery, resource rent would normally be dissipated completely (Gordon, 1954; Hannesson, 1993a). Table 4.4 indicates that a trawl-coastal fishery vessel combination of a little over 120–2500 vessels would dissipate discounted resource rent from the fishery to nil. Thus, this vessel combination or its equivalent is our model's prediction of the open access fishing capacity. If the agents in the fishery were to receive subsidies totaling NOK 9.48 billion (in present value), in addition to having open access to the resource, then the model's prediction of fishing capacity is 140 trawlers and 3000 coastal vessels or their equivalent. It should be interesting to compare the equilibrium stock and catch profiles that would result under "open access plus subsidy," open access, Nash non-cooperative, and the sole ownership equilibria. This is done graphically in Figures 4.1 and 4.2. The figures illustrate clearly the adverse effects of "open access plus subsidy," open access and Nash non-cooperative equilibria as compared to the optimal solution.

Table 4.3 Matrix giving the payoff to each player as a function of k_1 (no. of T vessels) and k_2 (no. of C vessels) in billions of NOK. Player T's payoff is placed in the southeast corner of the cell in a given row and column, and the payoff to player C is placed in the northeast corner

k1 (no. of T vessels)	*k2 (no. of C vessels)*											
	0	*500*	*700*	*900*	*1000*	*1050*	*1100*	*1200*	*1500*	*2000*	*2500*	*3000*
		28.22	31.76	35.34	35.99	35.79	36.11	35.3	33.89	28.63	26.67	20.87
0		0	0	0	0	0	0	0	0	0	0	0
	0	21.28	24.58	25.53	25.28	25.49	25.44	25.04	23.99	24.21	18.5	14.07
20	17.52	13.3	11.32	9.51	8.67	8.38	8.09	7.46	6.08	4.31	3.54	2.66
	0	15.54	17.1	17.7	17.67	18.13	17.99	17.89	17.12	14.47	13.3	9.85
40	28.72	18.8	15.29	12.75	11.65	11.39	10.92	10.07	8.14	5.64	4.48	3.19
	0	14.18	15.66	16.3	16.28	16.54	16.58	16.58	15.86	13.94	12.08	8.6
45	29.64	19.1	15.57	13.02	11.9	11.57	11.11	10.35	8.22	5.82	4.36	2.97
	0	12.92	14.48	15.02	15.09	15.42	15.28	15.4	14.66	12.99	11.4	7.54
50	30.57	19.25	15.89	13.23	12.04	11.78	11.19	10.47	8.23	5.74	4.27	2.66
	0	11.84	13.35	14.01	14.1	14.34	14.36	14.22	13.64	11.8	10.95	6.81
55	31	19.22	15.89	13.25	12.12	11.8	11.29	10.4	8.24	5.6	4.22	2.45
	0	11.48	12.98	13.66	13.66	14.03	13.99	13.76	13.54	11.73	10.46	6.45
57	31.45	19.2	15.97	12.1	12.1	11.84	11.37	10.36	8.32	5.59	4.34	2.26
	0	10.87	12.43	13.18	13.24	13.26	13.42	13.33	12.89	11.25	9.7	5.81
60	31.57	19.08	15.96	13.42	12.2	11.67	11.31	10.36	8.19	5.46	3.81	2.15
	0	10.29	11.74	12.3	12.7	12.67	12.66	12.75	12.11	10.81	8.81	6.01
65	32.17	19.4	16	13.3	12.31	11.74	11.29	10.42	8.04	5.35	3.6	1.95
	0	9.41	10.82	11.52	11.58	11.92	11.95	11.95	11.72	9.83	8.3	4.77
70	32.42	18.9	15.65	13.13	11.86	11.63	11.2	10.24	8.04	4.92	3.38	1.41
	0	8.13	9.46	9.88	10.26	10.67	10.56	10.67	10.39	8.24	7.47	3.58
80	31.7	18.2	15.06	12.24	11.39	11.34	10.67	9.83	7.54	4.17	2.71	0.54
	0	7.33	8.28	9.11	9.4	9.33	9.63	9.12	9.35	7.16	6.31	2.03
90	30.05	17.7	14.17	12.02	11.12	10.45	10.32	8.82	6.97	3.41	1.94	−0.63
	0	6.48	7.63	8.2	8.64	8.88	8.53	8.66	8.37	6.43	4.93	0.97
100	28.23	16.69	13.84	11.41	10.57	10.28	9.33	8.61	6.08	2.51	0.69	−1.65
	0	5.02	5.91	6.48	6.36	6.56	6.51	6.35	6.11	4.09	2.57	−0.69
120	23.5	14.03	11.24	9.1	7.73	7.38	6.84	5.79	3.62	0.15	−1.84	−3.89
	0	4.42	5.35	5.66	5.41	5.98	5.52	5.59	5.16	2.59	0.17	−2.74
140	21.63	12.89	10.42	7.81	5.99	5.66	5.08	4.42	1.99	−2.18	−4.58	−6.73

Table 4.4 Overall PV of economic rent from the fishery as a function of k_1 (no. of T vessels) and k_2 (no. of C vessels), in billions of NOK

	*k*2											
*k*1	*0*	*500*	*700*	*900*	*1000*	*1050*	*1100*	*1200*	*1500*	*2000*	*2500*	*3000*
0		28.2	31.8	35.3	36	35.8	36.1	35.3	33.9	28.6	26.7	20.87
20	17.5	34.6	35.9	35	34	33.9	33.5	32.5	30.1	28.5	22.1	16.73
40	28.7	34.3	35.4	30.5	29.3	29.5	29.8	27.9	25.3	20.1	17.8	13.4
45	29.6	33.3	31.2	29.3	28.2	28.1	27.7	26.9	24.1	19.8	16.4	11.57
50	30.6	32.2	30.4	28.2	27.1	27.2	26.5	25.9	22.9	18.7	15.6	10.2
55	31	31.1	29.2	27.3	26.2	26.1	25.7	24.6	21.9	17.4	15.2	9.26
57	31.4	30.7	29	26.8	25.8	25.9	25.4	24.1	21.8	17.3	14.8	8.71
60	31.6	30	28.4	26.6	25.4	24.9	24.7	23.7	21.1	16.7	13.5	7.95
65	32.2	29.7	27.8	25.6	25	24.4	24	23.2	20.2	16.2	12.4	.796
70	32.4	28.3	26.5	24.6	23.4	23.6	23.2	22.2	19.8	14.7	11.7	6.18
80	31.7	26.3	24.5	22.1	21.7	22	21.2	20.5	17.9	12.4	10.2	4.13
90	30	25	22.5	21.1	20.5	19.8	20	17.9	16.3	10.6	8.3	1.4
100	28.2	23	21.5	19.6	19.2	19.2	17.9	17.3	14.5	8.9	5.6	0.68
120	32.5	19.1	17.2	15.6	14.1	13.9	13.4	12.1	9.7	4.2	0.73	−4.59
140	21.6	17.3	15.8	13.5	11.4	11.6	10.5	10	7.2	0.4	−4.4	−9.48

Notice that contrary to expectations, the biomass is not completely depleted at the end of the game. There are two possible reasons for this. In the first place, players are not allowed to exploit all age groups of fish. Secondly, even if this was allowed, it would not be economically profitable to catch every single fish available.

Perfect malleability versus perfect non-malleability

The model in Chapter 3 is denoted the perfect malleable capital model and the model in this chapter is denoted the perfect non-malleable capital model. The purpose here is to compare the capacity investments and the PV of resource rent accruable to the players, both individually and to society at large, in the two models. To do this, the perfect malleable model is run using comparable prices and costs. Table 4.5 gives the capacity predictions of the two models, and the PV of resource rents to the players when both agents are active; when only T is active; and when only C is active. It is seen from this table that vessel capacity investment varies from year to year in the case of the malleable capital model.

A possible interpretation here is that each player evaluates his or her optimal capacity requirement in a given year and then rents precisely this quantity from a rental firm for fishing vessels.[13] For instance, when both are active, T's optimal vessel size varies from a high of 65 in the third year to a low of 41 trawlers in year 15, while C's varies from a high of 939 in year 3 to a low of 564 in the last year. In the non-malleable capital model, however, T's *ex ante* fixed capacity

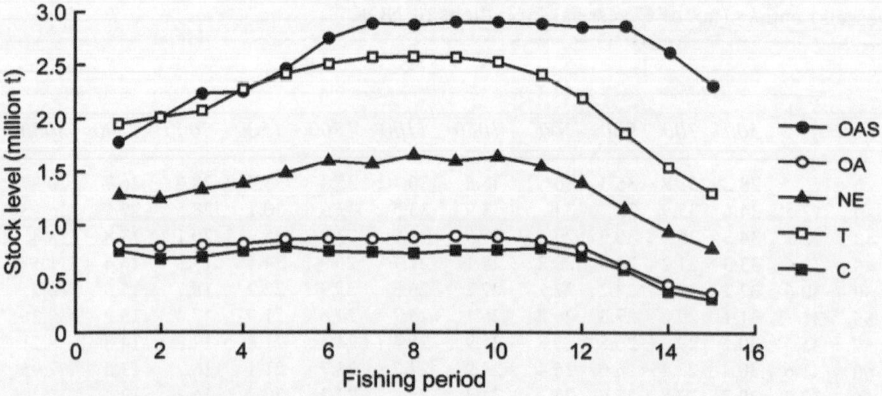

Figure 4.1 Stock profiles (million tonnes). Illustrating the post-catch stock size in each period for open access plus subsidy (OAS), open access (OA), Nash equilibrium (NE), T only and C only (the optimal solution).

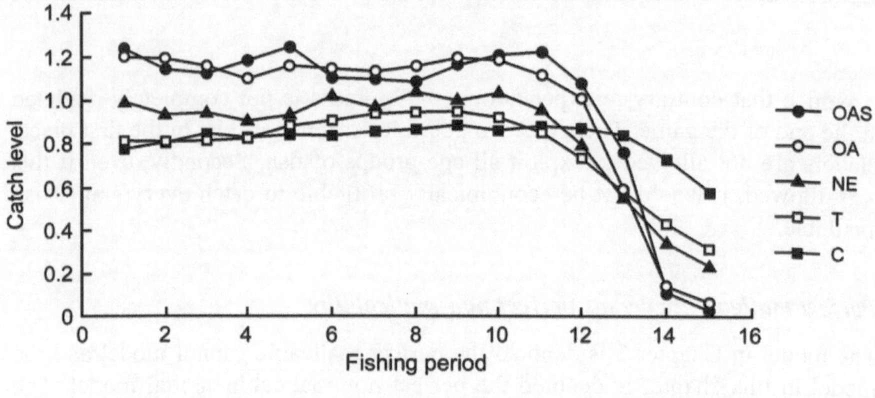

Figure 4.2 Catch profiles (million tonnes). Illustrating total catch in each period for open access plus subsidy (OAS), open access (OA), Nash equilibrium (NE), T only and C only (the optimal solution).

investment is 57 trawlers, and the corresponding capacity investment for C is 1050 coastal vessels.

The economic results given by the two models are given in the third column of Table 4.5. Two important observations can be made from this table. First, the maximum economic yields from the resource are different in the two models: NOK 44.53 billion is achieved in the malleable capital model and NOK 36.11 in the non-malleable one. The higher PV of resource rent achievable in the malleable capital model can be attributed to the removal of the restriction

Table 4.5 Malleable versus non-malleable capital giving the equilibrium vessel sizes and the overall discounted economic rent that accrues to society from the resource

	Vessel size (in numbers)		*PV of economic benefits (in billion NOK)*	
	Malleable	*Non-malleable*	*Malleable*	*Non-malleable*
Both active	(65–41; 939–564)	(57; 1050)	38.26	25.87
T active	(80–49)	(70)	42.35	32.42
C active	(1153–761)	(1100)	44.53	36.11

that non-malleability of capital imposes on the agents. The negative impact of this restriction is quite substantial, reaching up to about NOK 12 billion (or 47% of what is achievable under the restriction) in the case of the Nash equilibrium solution. This clearly demonstrates that there is much to be gained from establishing rental firms for fishing vessels, or rather, by allowing mobility of vessels between different stocks.[14]

Second, in both models the best economic results are achieved when C operates the fishery alone. However, the superiority of C becomes sharper in the non-malleable capital model: there is a difference of well over 10% (in favor of C) in the PV of resource rent accruable in the non-malleable capital model. In the malleable capital model, however, a difference of only about 5% is noted. This finding may have to do with the relatively high fixed costs of trawlers and the fact that fixed costs must be taken as given by the players in the non-malleable capital model.

Comparison with available data on the North-east Atlantic cod fishery

It is stated earlier that in 1991 the equivalent fishing capacity of about 638 coastal vessels and 58 trawlers was operated by Norwegian fishers to land 130 and 270 thousand tonnes of cod, respectively. This implies that catch per trawler is about 4655 tonnes and catch per coastal fishery vessel is 203 tonnes. Now, the Nash equilibrium strategies stipulate vessel sizes of 1050 for C and 57 for T. Together, these capacities are used to land an average of about 842 thousand tonnes a year.[15] Of the total, trawlers land 412 thousand tonnes and the coastal fishery vessels 430 thousand tonnes. Thus, catches per vessel are 7228 and 410 tonnes for a trawler and coastal fishery vessel, respectively. These numbers signify the incidence of overcapacity in the North-east Atlantic cod fishery even in comparison to the results from a non-cooperative solution. Comparison with the sole owner's optimal solution reveals an even greater degree of overcapacity in the fishery: in this case 1100 vessels are used to land an average of about 770 thousand tonnes of fish per year by C (that is 700 tonnes per vessel) and 70 trawlers land an average of about 780 thousand tonnes per year by T (well over 10,000 tonnes per vessel). A catch per trawl vessel of 10,000 tonnes per year appears to be high, probably the proportionality assumption (underlying the catch function) between the stock size and the catch per vessel is appropriate only when variation in the stock size is not too large.

Effect of fixed costs, discount rates, initial stock size, and terminal constraint

Fixed costs

Sensitivity analysis shows that (for the game solution), the elasticity of the PV of economic benefits accruable to the players T and C with respect to fixed costs is about −0.7 and −0.4, respectively.[16] Hence, to achieve the higher discounted resource rent, T needs a drop in its fixed costs relative to those of C of about 28%. The equivalent elasticities in the sole owner solutions are −0.3 for T and −0.2 for C, which implies that T needs a drop in its fixed costs relative to those of C of about 39% to take over from C as the producer of the optimal solution. The effects of zero fixed costs were also investigated. This is the same as assuming that fixed costs are considered to be "sunk" by the agents. Under such an assumption, C and T achieve discounted resource rents of NOK 20.27 and 19.7 billion, respectively, in the game situation, and NOK 42.06 (C) and 42.64 (T) in the sole ownership solutions. We see that the previously clear superiority of C is now neutralized to a great extent.

Discount rates

For the game situation, we found that a 1% drop in the *discount* rate faced by both players results in a 1% increase in the relative profitability[17] of T. On the other hand, an equivalent drop in the discount rate in the sole ownership scenario leads to an increase of 0.5% in the relative profitability of C. Intuitively, it is not difficult to understand why T does relatively better in the game situation while C does relatively better in the sole ownership case: Since T catches everything from age group 4 and above, while C catches only age groups 7 and above, it is no wonder that T is the one best positioned to capitalize on the increase in patience that a decrease in discount rate entails. C does better in the sole ownership scenario because an increase in patience plus the fact that C catches fish from age group 7 and above means that a larger proportion of the stock will reach maximum weight before it is caught, thereby resulting in better relative profitability for C.

Initial stock size

To investigate the effect of the initial stock size on the relative profitability of the agents, the model is re-run with 50% and 150% of the base stock size of 1.8 million tonnes. The results obtained indicate that in the sole ownership solutions, T improves its relative profitability as the stock size increases; from 86.85% when the stock size is only 50% of the base case to 92.1% when the stock size is 150% of the base case. The effect of stock size on the relative profitability of the agents in the game solution is however not that clear: T increases its relative profitability both when the stock size is only 50% of the original (90.6% as against 84.4% in the base case) and when the stock size is 150% of the base stock size. In this case, T's relative profitability is 92.3%. As these numbers show, T's relative profitability increases by a larger margin when the stock size increases than when it decreases.

Terminal constraint on the stock size to be left behind at the end of the game

A requirement that not less than 50% of the initial stock size should be left in the sea at the end of the game changes the outcome of the game significantly. In the sole ownership situation, however, the same solutions as in the base case are obtained, mainly because this constraint is not binding in these cases. Under such a requirement, it turns out that T comes out better (in contrast to the base case where it is C that does better), earning NOK 15.97 billion as against C's NOK 12.98 billion. An important point to note here is that the introduction of the terminal constraint, in the game environment, leads to an increase in the overall benefit from the fishery from NOK 25.87 to NOK 28.95 billion. This explains why economists advocate regulation when common property resources are exploited in non-cooperative environments.

Concluding remarks

The main findings of this study can be stated as follows: the optimal capacity investments in terms of number of vessels for T and C in a competitive, non-cooperative environment are 57 trawlers and 1,050 coastal vessels, respectively. The use of these capacities results in discounted benefits of NOK 11.84 and 14.03 billion, respectively, to T and C, and an overall discounted economic benefit of NOK 25.85 billion to society at large. Using only T and C vessels in the exploitation of the resource, the optimal fleet sizes are 70 and 1,100, respectively. In these cases the PV of resource rents are NOK 32.42 and 36.11 for T and C. It was also found that, as expected, the results obtained are rather sensitive to perturbations in fixed costs, discount rates, initial stock size and the terminal constraint.

It is in order to state here that, given that modeling and computation are always exercises in successive approximation (Clark and Kirkwood, 1979), our estimates should not be taken too literally. Having said this, the results of the analysis indicate that in its current state the North-east Atlantic cod fishery appears to suffer from over capacity. Hence, the practical implication of this study with respect to efficient management of the resource is that the excess capacity should be run down as rapidly as possible to a certain level,[18] somewhere above the

Nash equilibrium capacity level. Thereafter, the remaining excess capacity is allowed to depreciate to the "desired" level by itself. From then on, new capacity investment is undertaken only to make up for depreciation. This would ensure each player his or her best possible outcome and the society the second best solution.

The practical implication of the results obtained would have been somewhat different if the fishery were under-exploited – in this case, a distinction between starting a completely new fishery and the case where fishing is currently in progress. In the case of a new fishery, there is a real possibility for realizing the first best solution by allowing only C to exploit the resource with a capacity size of about 1,100 coastal fishery vessels. However, if political, social, and cultural realities dictate the participation of both players and this were to be in a non-cooperative environment, then each player should aspire to start off with its Nash equilibrium capacity size as computed herein.

5 Strategic dynamic interaction

The case of Barents Sea fisheries[1]

Introduction

The marine life in the Barents Sea supports two major fisheries – a fishery for cod and other demersal species, and a purse seine fishery for capelin.[2] It is known that there is a predator–prey relationship between cod and capelin. The purpose of this chapter is to study the economic effects of this biological interrelationship under different management arrangements. A model that captures this relationship is developed; thereafter a numerical method is applied to compute various equilibrium solutions of the model. First, Nash non-cooperative equilibrium solutions are determined when the two stocks are managed separately by their respective owners. Second, joint management equilibrium solutions are identified by assuming that exploitation and management of cod and capelin are carried out by a sole owner.[3] The latter solution is best in the sense that the sole owner is expected to internalize the externalities that are bound to originate from the natural interactions between the two species. Third, the exploitation of only cod is allowed in order to isolate the economic merits of allowing cod to feed on capelin while only cod is caught for human consumption. Note that cod is the more valuable of the two species.

Specifically, there are five main questions with which the chapter is concerned: (1) What is the maximum discounted resource rent that can be derived from the resource under joint and separate management? (2) How significant is the difference between these two solutions? (3) What is the effect of exploitation on the stock levels under these management regimes? (4) Is it economically optimal to exploit both species at current market conditions? (5) How are capelin catch and predation traded-off against each other, given changes in prices, costs, and discount factors?

A short historical note begins the chapter, and then a bioeconomic model for the cod-capelin fisheries of the Barents Sea is established. The data and numerical results are then presented. The solution procedure for the model is outlined in the Appendix.

Historical note

Quirk and Smith (1977) and Anderson (1975a) study and compare the free access equilibrium and the social optimum of ecologically interdependent fisheries. They derive necessary conditions for optimization and interpret these in general terms. Hannesson (1983) extends the results of these two papers by finding answers to the following questions: (i) to what extent are the results of single-species theory also valid for multispecies theory? (ii) Is there a well-defined relative price of products obtained at different levels in the food chain at which catching should be switched from one species to another? (iii) Will a stronger discounting of the future always imply a decreasing standing stock of biomass?

Three points should be noted about the above mentioned works. First, none of the papers analyse strategic conflicts and interactions. Second, as in Silvert and Smith (1977) and May *et al.* (1979), these papers use slightly different versions of the Lotka–Volterra model to characterize the multispecies systems they study. Hence, the implicit assumption in these papers is that the fish population is a homogeneous entity that can be adequately described by a single variable. Third, the papers are mainly theoretical, with very little or no empirical content; they are hence not applied to any specific fishery.

The fundamental game theoretic paper that analysed the problems associated with the joint management of fishery resources in detail is Munro (1979). The papers of direct bearing to this work are those of Fischer and Mirman (1992), and Flaaten and Armstrong (1991). These are theoretical analyses of interdependent renewable resources, which study game situations. In addition, these papers assume single cohort growth rules to derive general theoretical results. This chapter, by contrast, is on the one hand an empirical study of the Barents Sea fisheries, which explicitly recognizes that fish grow with time and that the age groups of fish are important both biologically and economically. On the other hand, a central aspect of this chapter is that it studies game situations and applies specific functions to analyse two specific fisheries. It is worth mentioning that these fisheries have previously been studied both biologically and economically for the purpose of finding the optimal rate of utilization of the resources therein (Eide and Flaaten, 1992; Hamre and Tjelmeland, 1982). Nevertheless, this study should provide further insights into the problems involved. For instance, in contrast to this paper, Eide and Flaaten (1992) analyse only the sole ownership outcome. These earlier multispecies bioeconomic models and analyses are precursors to recent analysis of the role of forage fish in the marine ecosystem (e.g. Hannesson *et al.*, 2009; Herrick *et al.*, 2009; Pikitch *et al.*, 2012).

The bioeconomic model

A multispecies system is modeled in which there are two biologically interdependent species. The interdependency derives from the fact that one of the species, cod, predates the other, capelin. This biological interaction is captured here through the way the weight of cod and the predation on capelin by cod are modeled. The simple assumptions made regarding these are (i) the weight of the predator (cod) is positively related to the density of capelin in the habitat; and (ii) the predation on capelin depends positively on the biomass of cod and the density of capelin in the habitat. It is also likely that the survival rate of cod will depend weakly on the abundance of good food, i.e. capelin. This effect is, however, considered negligible and thus ignored in this formulation. The formalization of these assumptions is based on Moxnes (1992), which in turn is inspired by the MULTSIMP model developed by Tjelmeland (1990).

Capelin

The capelin fishery takes place in two seasons (the winter and summer fishing seasons) of approximately two months duration each. The winter capelin fishery exploits mature capelin on its way to the spawning grounds. The most important age group exploited by this fishery is 4 years old, but some 3- and 5-year-olds are caught as well. The summer capelin fishery exploits fish of 2 years and above. In this model, the capelin fishery operate in the winter season only. The justification for this is threefold. First, the winter fishery is economically the more important of the two. Second, the winter fishery exploits mainly mature capelin, which are more valuable because they weigh more. Third, winter-caught capelin are more likely to last longer than summer capelin because they normally have less organic content in their diet. This is partly because a capelin stops feeding before spawning, after which it dies. It is assumed that all capelin mature at age four and confine the winter capelin fishery to age groups three and four. Hence, there are four capelin age groups in our model. At the opening of a fishing season, a constant number of 1-year-olds are recruited into the fishery. For a typical capelin cohort, a decrease in the stock comes from natural mortality, fishing mortality, and the predatory activities of cod on the cohort.

From now on the subscripts a ($a = 1, \ldots, A$) and t ($t = 1, \ldots, T$) represent age groups of fish (both capelin and cod) and fishing periods, respectively; and the superscripts co and ca refer to variables and parameters that relate to cod and capelin, respectively. Note that A and T denote the last age group and last fishing period, respectively.

The natural survival rate for capelin, s^{ca}, is assumed to be constant for all age groups. Fishing mortality is given by the catch function,

$$h_{a,t}^{ca} = q_a^{ca} e_t^{ca} \tag{5.1}$$

where the parameter q_a^{ca} is the age-dependent catchability coefficient; and e_t^{ca} is the fishing effort exerted on capelin in number of vessels (or fleet). The catch function is modeled in this manner because capelin is a schooling species; hence, the assumption is that once capelin schools are located, the fishing vessel is simply filled up in readiness for return to the port of call.[4]

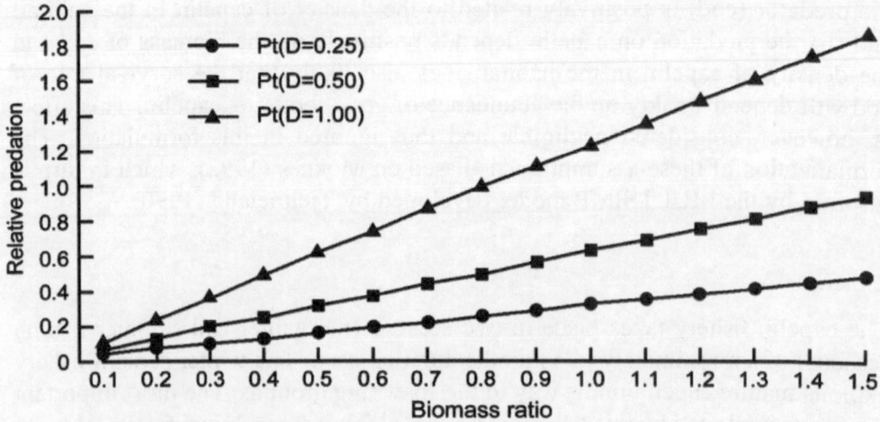

Figure 5.1 Relative predation versus biomass ratio at different levels of density of prey.

Following Moxnes (1992), let the predation on age group *a* capelin in period *t* by cod, $p^{ca}_{a,t}$, be given by

$$p^{ca}_{a,t} = \rho_t n^{ca}_{a,t} \tag{5.2}$$

Where $n^{ca}_{a,t}$ is the number of age *a* capelin in period *t*. ρ_t, denoting relative predation, is defined as the amount of capelin eaten by cod in period *t* divided by the total biomass of capelin in that period. Hence, $p^{ca}_{a,t}$, is the number of age group *a* capelin eaten by cod in period *t*.

The amount of capelin eaten is a function of both the biomass of the predator, B_{pred}, and the density of the prey, $D_{prey} \geq 0$. Furthermore, when $D_{prey} = 1$, each cod is assumed to eat k_1 times its own weight. Thus, ρ_t can be expressed as

$$\rho_t = \frac{k_1 D_{prey_t} B_{pred_t}}{B_{prey_t}} \tag{5.3}$$

Figure 5.1 illustrates how relative predation varies with changes in the biomass ratio (that is, biomass of predator divided by biomass of prey) at different prey densities. It is seen from equation (5.3) and Figure 5.1 that an increase in the biomass ratio results in an increase in relative predation, while an increase in prey density leads to an upward swing in the relative predation curve.

Note that when there is no capelin, $D_{prey} = 0$, ρ_t is also zero. The density of capelin in the habitat at time *t*, D_{prey}, is defined by the following equation (Moxnes, 1992):

$$D_{prey_t} = \frac{D^{\max}_{prey}}{1 + (D^{\max}_{prey} - 1)\left(\frac{B_{prey_t}}{B'_{prey}}\right)^{-\beta}} \tag{5.4}$$

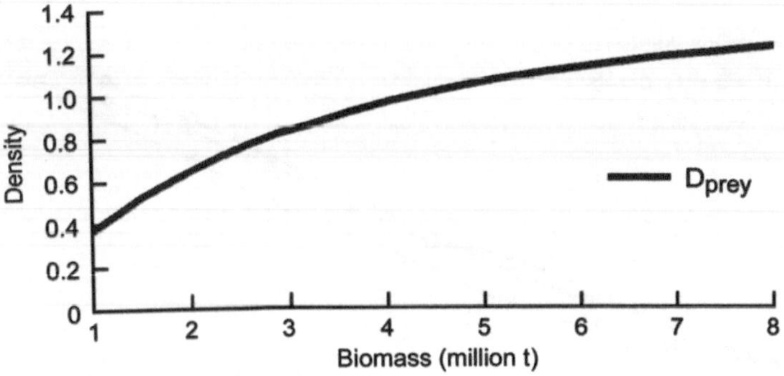

Figure 5.2 Density versus biomass of prey.

where B'_{prey} is a standard magnitude of capelin biomass at which $D_{prey} = 1$; D_{prey} is a constant maximum factor by which cod will increase its normal intake of capelin (given by k_1) at high densities of capelin; and $\beta > 0$ is a parameter. It is worth mentioning that the above relationship corresponds to the type 2 functional response reported in Holling (1965). Figure 5.2 illustrates the relationship between density of prey and prey biomass. As can be seen, the curve is concave.[5]

From the foregoing, the stock dynamics of the capelin stock can be represented by

$$
\begin{aligned}
&n_{1,t}^{ca} = R^{ca} \\
&n_{2,t}^{ca} \leq s^{ca} n_{1,t-1}^{ca} - p_{2,t}^{ca} \\
&n_{3,t}^{ca} + h_{3,t}^{ca} \leq s^{ca} n_{2,t-1}^{ca} - p_{3,t}^{ca} \\
&n_{4,t}^{ca} + h_{4,t}^{ca} s^{ca} n_{3,t-1}^{ca} - p_{4,t}^{ca}, n_{4,t}^{ca} \geq E, \quad \forall t \geq 1; n_{a,0}^{ca} \text{ given}
\end{aligned}
\tag{5.5}
$$

In the above equation, $n_{a,0}^{ca}$ denotes the number of age a capelin at the start of the game, R^{ca} is constant recruitment of capelin, and E represents the escapement required to maintain recruitment of the stock. This escapement is set equal to half a million tonnes (or about 20 billion individuals) as recommended by Hamre and Tjelmeland (1982). The stipulation of a minimum escapement implies that recruitment of capelin can be regarded as independent of the stock level so long as escapement does not occur below this threshold value.

Cod

In this case, a typical cohort of cod decreases in number due to only natural and fishing mortalities. But unlike in the case of capelin, where weight of individuals in a given age group is assumed to be constant, weight of cod is assumed to

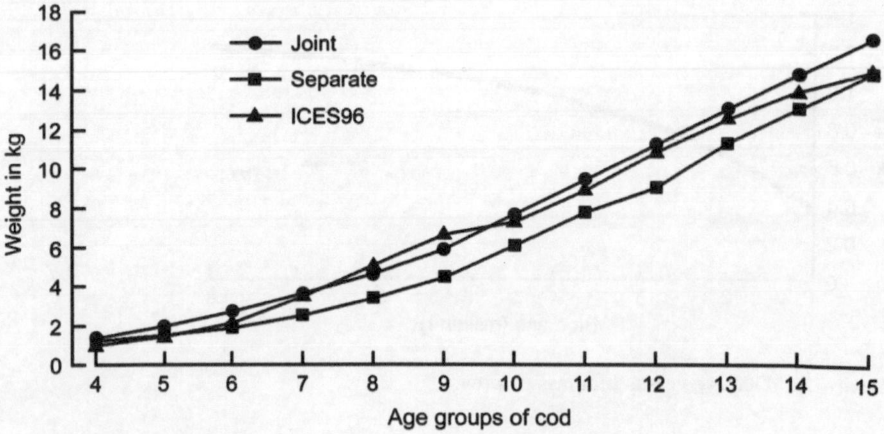

Figure 5.3 Weight versus age.

depend positively on the density of capelin, D_{prey}. The dependence of the weight of age group a cod in period t on the density of capelin is captured mathematically by (Moxnes, 1992):

$$w^{co}_{a,t} = w^{co}_{a-1,t-1} + GW_{a0}\left[D_{prey_t}k_2 + (1-k_2)\right], \quad w^{co}_{a,0} \text{ given} \tag{5.6}$$

Where $w^{co}_{a,0}$ is the weight of an individual in age group a cod (in kg) at the start of the game, GW_{a0} is the normal growth rate of age group a cod, and k_2 denotes the relative importance of capelin as food for cod in relation to other sources of nutrition. Notice that when there is no capelin in the habitat, D_{prey} is equal to zero, and the growth of cod would then depend only on other sources of nutrition given by the expression $GW_{a0}(1-k_2)$. See Figure 5.3 for a plot of the equilibrium weights given by equation (5.6) under separate and joint management.[6] For a given year-class of cod, the number of individuals decreases over time due to constant natural, and fishing mortalities, hence:

$$\begin{aligned}
&n^{co}_{0,t} = f(B^{co}_{t-1})\\
&n^{co}_{a,t} + h^{co}_{a,t} \leq s^{co} n^{co}_{a-1,t-1}, \quad \text{for } 0 < a < A, t > 0\\
&n^{co}_{A,t} + h^{co}_{A,t} \leq s^{co}(n^{co}_{A,t-1} + n^{co}_{A-1,t-1}), \quad n^{co}_{A,0} \text{ given}
\end{aligned} \tag{5.7}$$

where $f(B^{co}_{t-1}) = (\alpha B^{co}_{t-1})/(1+\gamma B^{co}_{t-1})$ is the Beverton–Holt recruitment function,[7] and $B^{co}_{t-1} = \Sigma_a p_a w^{co}_{a,t-1} n_{a,t-1}$ represents the spawning biomass in weight, p_a is the proportion of mature fish of age a; α and γ are constant biological parameters,[8] s^{co} is the constant survival rate of cod, $n^{co}_{a,t}$ represents the post-catch number of age group a cod in fishing period t. The catch of age group a

cod in fishing period t, is given by

$$h_{a,t}^{co} = q_a^{co} n_{a,t}^{co} e_t^{co} \tag{5.8}$$

where e_t^{co} is the fishing effort exerted on cod, and q_a^{co} stands for the age-dependent catchability coefficient of the cod catching vessels. Notice that in contrast to Eide and Flaaten (1992), where constant recruitment is assumed for cod, a recruitment function is specified for cod and constant recruitment for capelin assumed, mainly because capelin is a pelagic species, for which specifying a recruitment function is not an easy task.[9]

Economics

Non-cooperative (separate) management

Suppose there are two agents (i.e. the owners), each of whom catches only his or her own species. The fishery, hence, is organized under a cod and a capelin part, each managed by separate and distinct authorities. Organizing the fishery in this way can be justified both because cod and capelin are exploited by different parties using completely different technologies (trawlers for cod and purse seiners for capelin), and the fact that the fishing grounds of the two species are partly different. By this supposition and the fact that it is hard to imagine any market interaction between cod and capelin, we isolate externalities that arise only from the natural interactions between the two species (Fischer and Mirman, 1992).

The single period profit to the cod and capelin owners is derived from the sale of fish caught. These are defined as

$$\pi_t^{co} = v^{co} \sum_{a=4}^{A} w_{a,t}^{co}(D_{prey_t}) q_a^{co} n_{a,t}^{co} e_t^{co} - \frac{\psi^{co} {e_t^{co}}^{1.01}}{1.01} \tag{5.9}$$

for cod, and similarly

$$\pi_t^{ca} = v^{ca} \sum_{a=3}^{4} w_{a,t}^{ca} q_a^{ca} n_{a,t}^{ca} e_t^{ca} - \frac{\psi^{ca} {e_t^{ca}}^{1.01}}{1.01} \tag{5.10}$$

for capelin. Here, the subscripts and superscripts are as defined earlier: v denotes price per kilogram of fish caught, w represents weight, and ψ is the unit cost of hiring a given vessel type for one year. Notice that the single period profit to the cod owner depends on his or her own-effort and the stock size of both species. The dependence on the capelin stock stems from the weight of cod as this depends partly on the density of capelin. On the other hand, the single period profit to the capelin owner depends only on his or her own-effort, as weight of capelin is constant in this model.

Each owner is assumed to be interested in maximizing the sum of his or her discounted profit. Thus the cod owner maximizes

$$\Pi^{co} = \sum_{t=1}^{T} \delta^{co^t} \pi_t^{co} \tag{5.11}$$

with respect to both own-effort and own-stock level, subject to the stock dynamics given by equation (5.7), and the obvious non-negativity constraints. In equation (5.11), Π^{co} denotes the discounted sum of single period profits from cod, $\delta^{co} = (1 + r^{co})^{-1}$ is the discount factor, and $r^{co} > 0$ denotes the discount rate faced by the cod owner.

Similarly, the capelin owner maximizes

$$\Pi^{ca} = \sum_{t=1}^{T} \delta^{ca^t} \pi_t^{ca} \tag{5.12}$$

with respect to own-effort level, subject to the stock dynamics given by equation (5.5), and the obvious non-negativity constraints. Here, Π^{ca} denotes the discounted sum of single period profits from capelin, and δ^{ca} is the discount factor of the capelin owner. Notice that even though the stock level of capelin does not enter the profit function above, it does so in the constraints.

Joint management

Under sole ownership, the objective is to maximize the sum of the single period discounted profits from the two fisheries. Thus, the problem of the sole owner is to maximize

$$\Pi = \Pi^{co} + \Pi^{ca} \tag{5.13}$$

with respect to the effort levels exerted on the two species and their stock levels, subject to equations (5.5) and (5.7) above. In addition, it is understood that the obvious non-negativity constraints are satisfied. Here, Π denotes the discounted sum of single period profits from both cod and capelin.

Cod-only fishery

In this instance, the aim is to explore questions such as: Is there a relative price of cod or capelin at which it is economically sensible to catch only one of them? *A priori,* this question is relevant only in the case of a cod-only scenario. The capelin-only scenario is bound to give an inferior outcome relative to the case where both fisheries are active because of two reasons. First, no catching of cod would imply heavy predation on capelin, *ceteris paribus*. Second, capelin is the less valuable of the two species. Consequently, a capelin-only scenario is not modeled.

On the equilibrium solutions identified

We set out to identify Nash non-cooperative and sole ownership equilibrium solutions for the model outlined above. A Nash non-cooperative equilibrium in this context is a pair of strategy profiles, $\{(e_{ca}^*),(e_{co}^*, n_{co}^*)\}$, such that players will find it in their best interest to stick with their strategies if their opponents stick with theirs. On the other hand, an example of a joint management equilibrium is the outcome of the maximization of equation (5.13) under the relevant conditions.

Cavazzuti and Flåm (1992) show that not only do Nash equilibrium solutions exist in the two-person concave game formulated,[10] the equilibrium tends to be unique if profile players face the same shadow prices along the equilibrium.

It should be noted that the solutions computed in the non-cooperative scenario do not subscribe fully to the customary open loop solution concept derived from control theory. Unlike here where agents impact on their rival's stock indirectly through their choice of effort level, in the customary open loop solutions, agents are expected to directly control their rival's stock once the rival has committed to a given profile of actions.

Numerical results

To solve the model, a numerical procedure is applied whose mathematical formulation is developed in Flåm (1993), and applied to solve the single species model in Chapters 3 and 4. The parameters used for the computations are given in Table 5.1. The data comes from a number of sources including, the Institute of Marine Research (1994), ICES (1992, 1996), Kjelby (1993), Moxnes (1992), Flåm (1994), and Digernes (1980). Note that for the sake of scaling, a fleet size of 10 trawlers and 10 purse seiners are used as the unit of fishing effort. The simulation runs for the next 20 years.

The results

A plot of weight versus age of cod for a typical year-class given by our model under joint and separate management is given in Figure 5.3. In addition, a plot of the weight of the different age groups of cod reported in ICES (1996) is given on the same graph. We see from the graph that (i) joint management produces cod with the most weight, especially for older age groups; (ii) non-cooperation produces cod with the least weight; and (iii) current ICES estimates of the weight of cod lies in between those for the joint and separate management cases, which shows that the current effort at joint management of the two species is yielding some positive results, although short of what our model predicts.

Payoffs under the different management regimes

Table 5.2 presents the payoffs to the agents under the different management regimes. Column 2 of Table 5.2 gives the base case outcomes, while columns 3 to 5 present the outcomes from (*ceteris paribus*) sensitivity analysis.

Table 5.1 Parameter values used in the model

Biological parameters		*Comments/source*
$s^{co} =$	0.81	Standard for cod
$s^{ca} =$	0.535	Eide and Flaaten (1992)
$\alpha =$	1.5 per million tonnes	Chosen to give unfished biomass of 5 million tonnes
$\gamma =$	1.0 per million tonnes	Chosen to give unfished biomass of 5 million tonnes
$R^{ca} =$	500 billion individuals	cf. Digernes (1980)
$E =$	0.5 million tonnes	Tjelmeland (1982)
$\beta =$	1.2	Choice as in Moxnes (1992)
$k1 =$	1.235	Choice as in Moxnes (1992)
$k2 =$	0.6	Choice as in Moxnes (1992)
$B'_{prey} =$	4.467	Choice as in Moxnes (1992)
$D^{max}_{prey} =$	1.5	Choice as in Moxnes (1992)
GW_{a0}	(0.2, 0.21, 0.25, 0.3, 0.35, 0.45, 0.562, 0.744, 0.826, 1.0, 1.4, 1.4, 1.45, 1.45, 1.45, 1.5)	Based on data in Moxnes (1992)
$n^{co}_{a,0} =$	(0.46, 0.337, 0.298, 0.223, 0.117, 0.061, 0.033, 0.009, 0.009, 0.009, 0.009, 0.009, 0.009, 0.009, 0.009) in billion numbers	Average of initial numbers from 1984 to 1991 reported in Table 3.12 of ICES (1992)
$n^{ca}_{a,0} =$	(500,240,163,78,38) in billion numbers	Choice based on data in IMR (1994)
$p_a =$	(0, 0, 0, 0, 0, 0, 0, 1, 1, 1, 1, 1, 1, 1, 1)	Knife-edge selectivity applied
$w^{ca}_a =$	(0.012, 0.018, 0.021, 0.022)	Moxnes (1992) in kg
$w^{co}_a =$	(0.3, 0.6, 1, 1.4, 1.83, 2.26, 3.27, 4.27, 5.78, 7.96, 9.79, 11.53, 13.84, 15.24, 16.34)	ICES (1992) in kg
Economic parameters		*Comments/source*
$v^{co} =$	NOK 6.78 per kg	Sumaila (1995)
$v^{ca} =$	NOK 0.6 per kg	Moxnes (1992)
$k^{co} =$	NOK 210 million	Kjelby (1993)
$k^{ca} =$	NOK 10 million	Based on data in Flåm (1994)
Interest rate =	7%	Recommended by Ministry of Finance, Norway
Technological parameters		*Comments/source*
$q^{ca} =$	0.175	Based on data in Moxnes (1992)
$q^{co} =$	0.068	Sumaila (1994)

Table 5.2 Payoffs from cod and capelin under different management regimes (in billion NOK)

Management regime			*25% increase in price*		*25% increase in cost*		*Increase (0.935 to 0.99) in DF*[a]	
		Base case	*Cod*	*Capelin*	*Cod*	*Capelin*	*Cod*	*Capelin*
Joint	Cod	64.42	84.73	64.31	59.27	64.46	–	–
	Capelin	2.77	2.83	3.93	2.68	2.71	–	–
	Total	67.19	87.56	68.24	61.95	67.17	–	–
Separate	Cod	48.56	62.42	46.89	44.80	50.44	76.86	44.92
	Capelin	5.88	6.59	7.78	5.78	5.63	7.31	9.60
	Total	54.44	69.01	54.67	50.58	56.07	84.17	54.52
Cod only		65.58	85.98		60.48		103.70	

[a] DF, discount factor.

From column 2, we see that, as expected the best economic result of Norwegian Kroner (NOK) 67.19 billion (capelin contribution 4%) is achieved when both species are caught under joint management. On the other hand, the worst economic result of NOK 54.44 billion (capelin contribution 11%) is obtained when the species are caught under separate management. A situation where only the cod fishery is active yields a result (NOK 65.58 billion) better than that obtained under separate management, but worse than that under joint management.

The economic loss stemming from the externalities that arise due to the natural interactions between the two species is significant, reaching up to NOK 12.75 billion, or about 23% of what is achievable under separate management. The higher benefits accruable under joint management are due to a sensible allocation of the prey stock between predation and fishing. A good part of the capelin stock is not fished in the joint management case but rather left for the cod species to feed on. For instance, Table 5.4 reveals that a total of only 7.58 million tonnes of capelin is fished under joint management. Compare this with the 15.93 million tonnes caught under separate management, and the point made here will immediately become clear. Also, it should be noted that, as shown in Table 5.4, catch plus production of capelin is higher under separate management, which partly explains why the average annual standing biomass of capelin is higher under joint management.

An increase (decrease) in the price (fishing cost) of cod results in an increase (decrease) in the respective payoffs from cod and capelin under both separate and joint management (column 3 and 5). This is because an increase (decrease)

Table 5.3 Average annual standing biomass and yield under the two management regimes (in million tonnes)

Management regime	*Cod*		*Capelin*	
	Stock	*Catch*	*Stock*	*Catch*
Joint	2.23	1.24	1.55	0.38
Separate	2.62	0.94	0.90	0.80

Table 5.4 Effect of changes in economic parameters on capelin catch and predation (in million tonnes)

Management regime		*Base case*	*25% increase in price*		*25% increase in cost*		*Increase (0.935 to 0.99) in DF*[a]	
			Cod	*Capelin*	*Cod*	*Capelin*	*Cod*	*Capelin*
Joint	Catch	7.58	7.73	9.10	7.48	7.34	–	–
	Predation	10.31	9.50	10.40	10.31	10.42	–	–
Separate	Catch	15.93	18.97	16.90	15.17	15.48	22.37	18.14
	Predation	7.45	4.81	6.87	7.96	7.65	3.12	6.35

[a] Tables 5.3 and 5.4 do not include the cod-only scenario because the purpose here is to reveal the trade-off between catch and predation in the separate and joint management regimes.

in the price (fishing cost) of cod results in higher fishing mortality on cod, which in turn releases more capelin for fishing.

A positive change in the price of capelin leads to a decrease in the payoff to the cod owner, and an increase in the payoff from capelin under both separate and joint management (column 4). Such an increase in price makes it economically sensible to catch more capelin, thereby making less capelin available for predation. The opposite results are obtained with an increase in the fishing cost of capelin. The interesting point here is that under separate management, the gain in payoff by the cod owner is higher than the loss in payoff to the capelin owner, so that overall, an increase in the cost of fishing capelin by 25% leads to an increase in the total payoff to the fishing community.

An increase in the discount factor faced by one or the other of the two fisheries in the non-cooperative scenario leads both to an increase in the total payoff from the resource, and the share or contribution of the fishery facing the increase. Also, allowing only the cod fishery to be active tends to be more plausible as the cod fishery faces a relatively higher discount factor than the capelin fishery.[11]

Stock sizes, catches, and predation level

Generally, the computed outcomes confirm the results of Fischer and Mirman (1992). From Table 5.3, we see that joint management leads to a lower average annual catch of capelin (0.38 million tonnes) and a higher average annual catch of cod (1.24 million tonnes) compared to the separate management case at 0.8 and 0.94 million tonnes, respectively. Also, the average annual standing biomass of 3- and 4-year-old capelin turns out to be higher under joint management (1.55 million tonnes) than under separate management (0.9 million tonnes), while the average annual standing biomass of cod is lower under joint management (2.23 million tonnes) than under separate management (2.62 million tonnes). A probable explanation of the latter result is that the higher growth rate of cod implied by joint management means that sustainable catches of cod are achievable at a lower standing biomass.

Table 5.4 presents the total capelin catch and predation for increased prices, costs, and discount factors. Table 5.4 reveals that, under separate management, an increase in the price of cod or a decrease in its catch cost leads to an increase in the catch of capelin and a decrease in predation by cod; and an increase in the price of capelin or a decrease in the cost of catching capelin also leads to a decrease in predation and an increase in catch of capelin. It can also be seen from Table 5.4 that an increase in the discount factor of either fishery results in an increase in the catch, and a decrease in the predation of capelin. The intuition behind these results has been outlined in the previous section.

Concluding remarks

This study shows that there will be an economic loss if cod and capelin are exploited as if there were no biological interaction between them. Allowing for the fact that modeling and computations are exercises in successive approximations, this loss is computed to be nearly 25% of what is achievable if this interaction is neglected. In the summer of 1992 and the winter of 1993, 0.2 and 0.57 million tonnes of capelin were landed, respectively, from the Barents Sea (Institute of Marine Research, 1994, table 1.5.1). This means that a total of 0.77 million tonnes of capelin was landed in the 1992–1993 fishing year. Our model gives an average annual catch of 0.8 and 0.38 million tonnes of capelin, respectively, under non-cooperation and cooperation. This is an indication that current management practice does better than what would be achieved under non-cooperation, but clearly, it does not leave enough capelin to be "fished" by cod, as would be necessary under cooperative management.

Two other studies (Flaaten, 1988; Eide and Flaaten, 1992) come to similar conclusions. This would tend to make a strong case for a severe curtailment of the capelin fishery in the Barents Sea. It is important to highlight the fact that the biological models applied in these studies do not perfectly capture the predator–prey relationships between cod and capelin, not to mention the fact that these studies are partial in the sense that they do not include all the important predators (seals and whales) and preys in the habitat. In addition, this study is deterministic and thus, cannot be expected to give a perfect picture of the world under investigation. Nevertheless, the results of this chapter should give the relevant fisheries managers some food for thought.

6 Cannibalism and the optimal sharing of the North-east Atlantic cod stock[1]

Introduction

Over the years, fisheries managers in many countries have come to accept the concept of bioeconomic management of fish resources. The application of bioeconomics has usually been limited to the determination of total allowable catch (TAC), while the sharing of the TAC between heterogeneous fisher groups has been thought to be an equity issue, subject to political negotiations. The fact that different fisher groups catch different cohorts within fish stocks, and thereby have different effects upon both stock growth and the economics of the fishery, is not taken into account. Political determination of catch shares has bioeconomic effects, for instance, in the shape of reduced payoffs from the fishery, or even extinction. The bioeconomic losses could become quite substantial when there is cannibalistic interaction between sub-stocks within a single species. In this chapter, cannibalistic interactions between two sub-stocks that are fished by two separate vessel groups are studied. We show how, in the same manner that bioeconomics has become an important tool in multispecies management, a similar approach can be used profitably to manage species with intra-stock interaction.

A bioeconomic model is developed in order to study optimal[2] catch shares for two vessel groups, namely, trawlers and coastal vessels.[3] This is done using a cooperative game theoretic approach to the issue of sharing the catch of the North-east Atlantic cod stock. Russia and Norway, the former country relying solely upon trawler technology, jointly manage this stock.[4] These two countries meet annually to decide the TAC, and their respective shares of this catch. Furthermore, Norway must determine how to divide its share of the TAC between heterogeneous fisher groups; that is, trawlers and coastal vessels, fishing different cohorts within the stock. The coastal vessels target mainly mature cod, while the trawlers catch immature cod. In this chapter, the joint (cooperative) and separate (non-cooperative) outcomes to this resource-sharing problem in a cannibalistic interaction model were explored.

Spulber (1985) introduces the problem of using lumped parameter models for non-selective fishing decisions in multi-cohort stocks. He shows how sustainable catch within a biomass model may lead to actual extinction if catch is concentrated on recruits or spawners. His analysis is purely theoretical, illustrating the problem by allowing a sole owner to catch either selectively or non-selectively. Chapter 5 presents a multi-cohort model for the analysis of the management of the North-east Atlantic cod stock. However, this model does not explicitly allow for cannibalistic behavior which is presented, and which Eide (1997) shows to be an important explanatory function of changes in the North-east Atlantic cod stock. Armstrong (1999) studies cannibalism and the sharing of catches, but does not allow for optimal fishing in the build-up phase of the stock, such as is allowed in the current model. Klieve and MacAulay (1993) analyse the southern bluefin tuna (*Thunnus maccoyii*) fishery, where Japanese and Australian fishers catch different cohorts within the stock. The authors define different catch strategies for the two countries with respect to the choice of age at catch. Klieve and MacAulay (1993) determine which strategy combinations give the highest joint payoff to the players, by applying the Nash (1953) bargaining solution concept (Munro, 1979). This approach differs from the one in this chapter in that the current analysis does not limit the study to the cooperative solution given by Nash (1953), which in itself gives preference to one country when there is asymmetry between the countries. It is furthermore assumed that the catch strategies of the two vessel groups are determined by their existing technologies and their respective fishing grounds. Thus, the overall optimal sharing of the resource can be determined, after deciding the weights that should be given to the two parties' preferences.

Comparisons with the results in Chapters 3 and 4, where cannibalism is not taken into account, show that cannibalistic interaction can result in large economic losses. The incorporation of cannibalistic behavior also affects how the optimal annual catch should be shared between the coastal and trawler fleets. Results from the analysis shows that fishing by both fleet types is bioeconomically superior to a corner solution, that is, an outcome in which only one of the two vessel groups takes the whole TAC, in contrast to the results in Chapters 3 and 4. Furthermore, it was found that the existing allocation rule for the North-east Atlantic cod, which determines the shares of the TAC allotted to the two vessel groups, gives a sub-optimally high share of the total catch to the trawler fleet. In addition, the absence of cooperation between the two vessel groups leads to stock levels well below the safe minimum recommended by biologists (e.g. Jakobsen, 1993).

The North-east Atlantic cod and its fishery are presented in the next section. Following this is a section in which the bioeconomic model of the fishery is described. The results of the simulations are then presented, followed by a discussion and conclusion.

The North-east Atlantic cod fishery

The North-east Atlantic cod stock is highly migratory, working its way through both Norwegian and Russian "exclusive economic zones," as well as international waters. Norway and Russia together annually determine the TAC, giving each country approximately 45% of the TAC, with the remainder caught by other countries, such as Iceland, the Faroe Islands and some EU countries. The Russian and other country catch is mainly fished by trawlers offshore, while the Norwegian share of the TAC (the NTAC) is divided between two vessel groups: trawlers and coastal vessels. Since 1990, this division has been determined by a rule called the trawl ladder.[5] Based on the recommendation of the Norwegian Fisher's Association, which represents both vessel owners and fishers, the Norwegian government chose to implement the trawl ladder allocation rule. This rule is applied to calculate the shares of the catch of the two vessel groups depending on the size of the NTAC. The rule stipulates that a minimum trawler share of 28% should be allocated when the NTAC is below 130,000 tonnes. For higher NTACs, the trawler share increases linearly, with a maximum trawler share of 33% when the NTAC reaches 330,000 tonnes. Since almost all the non-Norwegian catch is taken using trawl gear, this means that the total trawler share is approximately 70% when the TAC for the North-east Atlantic cod stock is large. This can then be compared to the optimal shares derived from our model.

The model

A deterministic bioeconomic model, with two agents targeting separate but interacting sub-stocks within a fish stock, is presented. The two sub-stocks consist of different age groups, with mature fish in one sub-stock and immature fish in the other. The two sub-stocks interact via cannibalism and recruitment, with the mature preying upon the immature, and the immature recruiting to the mature sub-stock when reaching maturity. Fisher interactions are modeled in a dynamic game-theory setting, allowing us to study both cooperative and non-cooperative behavior (see, for example, Munro, 1979).

We concentrate on a single-stock version of the two-stock model presented by Lotka (1925) and Volterra (1928). Eide (1997) shows that this structure has a close fit to the biological findings regarding the changes in the North-east Atlantic cod stock throughout the 1980s. The changes in the biomass levels of the two sub-stocks are described by the following difference equation:

$$\Delta x_{i,t} = G_i(x_{1,t-1}, x_{2,t-1}) - h_{i,t}, i = 1, 2 \tag{6.1}$$

where $\Delta x_{i,t-1} = x_{i,t} - x_{i,t-1}$ and $x_{i,t}$ is the biomass of sub-stock i at time t, with $i = 1, 2$ defining immature and mature sub-stocks, respectively. It should be noted that $x_{i,t}$ can be expressed in terms of weight and number of fish to obtain $x_{i,t} = w_i * n_{i,t}$ where w_i is the average weight of sub-stock i and $n_{i,t}$ denotes the number of sub-stock i cod in period t. The rate of catch of sub-stock i, is defined as $h_{i,t} = q_i x_{i,t} e_{i,t}$, where q_i is the catchability coefficient of vessel group i, and $e_{i,t}$ is the number of vessels fishing the ith sub-stock in period t.

The natural growth functions G_i, of sub-stocks 1 and 2 also define the interaction between the two sub-stocks, and may be described as follows

(Eide 1997)[6]:

$$G_1(x_{1,t}x_{2,t}) = r_1 x_{1,t}\left(\frac{1 - x_{1,t}}{\alpha_1 x_{2,t}}\right) - bx_{1,t}x_{2,t}$$

$$G_2(x_{1,t}x_{2,t}) = r_2 x_{2,t}\left(\frac{1 - x_{2,t}}{\alpha_2 x_{1,t}}\right) \tag{6.2}$$

The parameters r_i, α_i and b are positive constants, with r_i being the intrinsic growth rate of sub-stock i.[7] The parameter b determines the cannibalistic interaction, where the size of sub-stock 1 is negatively affected by that of sub-stock 2. By putting x_j into the growth function $G_i(i \neq j)$, as described in the bracketed terms in equation (6.2), a recruitment relationship between immature and mature fish is allowed.

It is assumed that two different fishing agents or vessel groups (i.e. trawlers and coastal vessels designated 1 and 2, respectively), each catch their respective sub-stocks, 1 and 2. Hence vessel group i only catches sub-stock i.[8] Let the cost function of a given vessel type i in period t, $C(e_{i,t})$, be defined as in Chapters 3 and 4:

$$C(e_{i,t}) = \frac{k_i e_{i,t}^{1+\omega}}{1+\omega} \tag{6.3}$$

where $\omega = 0.01$, and $k_i/(1+\omega) \approx k_i$ is the cost of engaging one fishing fleet (or vessel) for one year. The above formulation introduces concavity in the objective function (necessary to ensure convergence: Flåm 1993) while still maintaining an almost linear cost function.[9] Hence, the single period profit of vessel group i = 1, 2 can be expressed as:

$$\pi_{i,t} = \pi_i(x_{i,t}, e_{i,t}) = v_i h_{i,t}(x_{i,t}, e_{i,t}) - C(e_{i,t}) \tag{6.4}$$

where v_i is the price per unit weight of sub-stock i. The stream of discounted single period profits of a vessel group, M_i, i = 1, 2, is defined as:

$$M_i(x_i e_i) = \sum_{t=1}^{T} \delta_i^t \pi_i(x_{i,t}, e_{i,t}) \tag{6.5}$$

where $\delta_i = (1 - r_i)^{-1}$ is the discount factor, and i denotes the discount rate of player i. Note that $t = (1, \ldots, T)$ represents fishing periods, with T denoting the end period.

Under a cooperative regime, the goal of the cooperative agents is to find a sequence of effort, $e_{i,t}$ and sub-stock levels, $x_{i,t}$, i = 1, 2, to maximize a weighted average of their respective objective functionals (that is, their stream of discounted single period profits):

$$\prod(x_1, x_2, e_1, e_2) = \beta M_1(x_1, e_1) + (1 - \beta) M_2(x_1, e_2) \tag{6.6}$$

subject to the stock dynamics given by equation (6.2) above, and the obvious non-negativity constraints. β and $(1-\beta)$ indicate how much weight is given to the objective functionals of 1 and 2, respectively, in the cooperative management problem. The following modified Lagrangian function can be set up for this problem (see Appendix):

$$L(x_1, e_1, x_2, e_2; y) = \prod(x_1, x_2, e_1, e_2) + y\phi^-(x_1, x_2, e_1, e_2) \tag{6.7}$$

where y is a modified Lagrangian in the sense of Flåm (1993):

$$y\phi^-(x_1, x_2, e_1, e_2) := \sum_{t=1}^{T} \begin{bmatrix} y_{1,t}H(G_1(x_{1,t-1}, x_{2,t-1}) - h_{1,t} - \Delta x_{1,t}) \\ +y_{2,t}H(G_2(x_{1,t-1}, x_{2,t-1}) - h_{2,t} - \Delta x_{2,t} \end{bmatrix}$$

and

$$H(G_i(x_{i,t-1}, x_{j,t-1}) - h_{i,t} - \Delta x_{i,t}) := \begin{cases} 1 & (G_i(x_{i,t-1}, x_{j,t-1}) - h_{i,t} - \Delta x_{i,t}) < 0 \\ 0 & \text{otherwise} \end{cases}$$

Under a non-cooperative regime, the problem of player i is to find a sequence of effort $e_{i,t}$ and own sub-stock $x_{i,t}$ ($t = 1, 2, \ldots, T$) to maximize his or her own objective functional denoted by M_i, subject to the relevant constraints. For this problem, the following modified Lagrangian function for each player can be formulated as:

$$L_i(x_i, e_i, x_j, e_j, y) = M_i(x_i, e_i) + \sum_{t=1}^{T} [y_{i,t}H(G_i(x_i, x_j, e_i, e_j) - h_{i,t} - \Delta x_{i,t})]$$

$$\forall i \neq j \tag{6.8}$$

The key difference between the cooperative and non-cooperative scenarios is that in the latter, each player maximizes without due regard to the intra-stock interaction between the two sub-stocks.

The solutions to equations (6.7) and (6.8) are pursued numerically using the solution procedure developed in Flåm (1993). From these solutions, the stock sizes under cooperative and non-cooperative interaction between the two vessel groups are obtained. Furthermore, the optimal equilibrium catch of the two sub-stocks (and by default the effort profiles underlying these catch levels) can also be determined. Summing these for each t the total optimal equilibrium catch in each time $t$is obtained.[10]

The parameter values used in the simulations are based entirely on data from Norwegian fishing vessels. The effort e_i, $i = 1, 2$, denotes the number of vessels within each vessel group. Hence, the economic parameters k_i, q_i and v_i are given the values in Table 6.1. Foreign trawlers are assumed to face the same economic and biological constraints as the Norwegian trawlers.

The discount factor δ is set equal to 0.935, as prescribed by the Norwegian Ministry of Finance, while b is found by Eide (1997) to be 0.2023.[11]

Table 6.1 Economic and biological parameter values (q, the catchability coefficient, is a per vessel value; k, the cost parameter, is measured in 10^6 NOK per year; while v, the price, is in NOK/tonne, x_0, the initial stock size, is in thousand tonnes). Vessel group 1 consists of trawlers, while 2 describes the coastal vessels

	Sub-stock/vessel group i	
	1	*2*
r^1	0.5003	0.6728
a^1	8.7608	1.1880
q^2	0.006650	0.001175
k^3	18.602103	1.452341
v^3	7,579	8,655
x_0	783,900	280,500

[1] r, the intrinsic growth rate, and a, the growth parameter are determined in Eide (1997).
[2] The catchability coefficients q, are average values decided by the actual catches, the vessel numbers (Anon., 1990b,1991, 1992, 1993), and the resulting stock sizes in the years 1990–1993.
[3] The cost parameters are given by the weighted (with regard to number of vessels and year) cost data in Anon. (1990b, 1991, 1992, 1993). The price parameters are the average prices that the two vessels obtained in 1992 (data from the Directorate of Fisheries).

Simulation results

Discounted resource rent

Table 6.2 presents the discounted profits to the coastal and trawler vessel groups, for different management preferences. The following points can be made. First, the best total economic result (over 25 years) is NOK 30.71 billion obtained when β (which denotes the weighting of preferences of the trawler fleet) is equal to 0.6. Of this, NOK 13.35 and 17.36 billion are obtained from the trawl and coastal fleets, respectively. Second, under non-cooperation, the potential economic benefits are wasted almost completely, with the coastal fleet and trawlers making NOK 1.47 and 1.37 billion, respectively. Third, as β approaches 0 or 1, total rent declines, an indication that allowing only the trawl or coastal fleet to exploit cod does not produce superior outcomes. The latter two results are in contrast to the results produced in Chapter 3.

Catch proportions and effort profiles

In Table 6.3, the average catch over the 25-year-period is presented for both trawlers and coastal vessels, given different management preferences. The results reinforce the observations made regarding resource rent in the previous section. We observe from Table 6.3 that the optimal annual catch is computed to be 450,000 tonnes (when $\beta = 0{:}6$), with the trawlers landing an average of 198,000 tonnes, and the coastal fleet 252,000 tonnes. This implies that the highest

Table 6.2 Discounted profits in billion NOK (present value over 25 years), for $0 < \beta < 1$, and for the non-cooperative outcomes. Numbers in bold indicate the profits that ensure maximum economic rent. Recall that β refers to the preferences of the trawl fleet

	Profit		
β	*Trawl*	*Coastal*	*Total*
0.1	2.39	9.40	11.79
0.2	4.19	2.29	13.48
0.3	7.93	11.80	19.73
0.4	11.27	14.95	26.22
0.5	12.89	16.94	29.83
0.6	**13.35**	**17.36**	**30.71**
0.7	13.14	15.27	28.41
0.8	14.18	7.85	22.03
0.9	14.22	3.05	17.27
Non-cooperative	1.37	1.47	2.84

Table 6.3 Average catch in million tonnes (over 25 years), for $0 < \beta < 1$, and for the non-cooperative outcomes. Numbers in bold indicate the catch/catch share that ensure maximum economic rent

	Average catch			
β	*Trawl*	*Coastal*	*Total*	*Trawl %*
0.1	0.0337	0.1550	0.1887	17.9
0.2	0.0667	0.1710	0.2377	28.1
0.3	0.1200	0.2100	0.3300	36.4
0.4	0.1670	0.2520	0.4170	39.9
0.5	0.1910	0.2660	0.4570	41.8
0.6	**0.1980**	**0.2520**	**0.4500**	**44.0**
0.7	0.1980	0.2080	0.4060	48.8
0.8	0.2270	0.1010	0.3280	69.2
0.9	0.2380	0.0386	0.2766	86.0
Non-cooperative	0.0361	0.0420	0.0781	46.2

discounted total profit is achieved when about 44% of the catch is taken by the trawl fleet.

Table 6.3 also shows that non-cooperative behavior leads to disastrous outcomes as all the catch potential is virtually wiped out: the annual average catch is only 78,100 tonnes.

In Figure 6.1, the catch profiles resulting from cooperative and non-cooperative fishing are presented for the 25-year period. We observe that the optimal catch

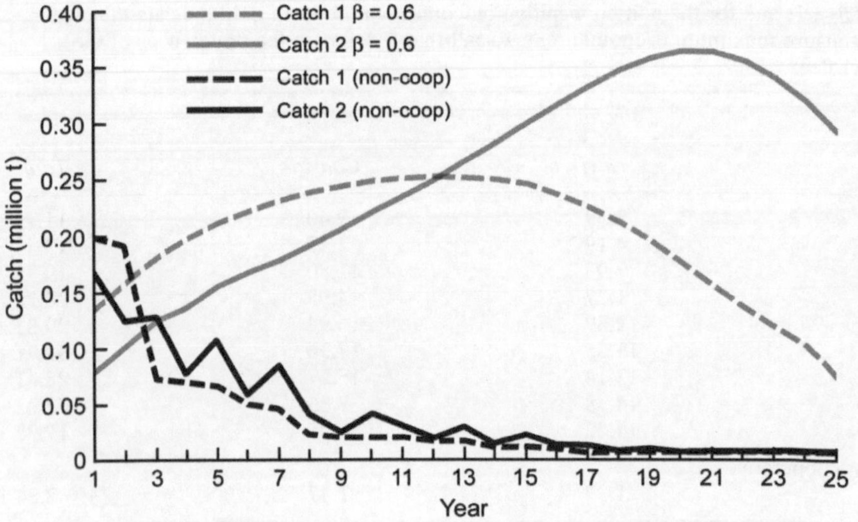

Figure 6.1 Catch profiles over a 25-year time period for the optimal cooperative and non-cooperative cases. Note that catch 1 and catch 2 refer to trawl and coastal fleet catches, respectively.

for both vessel groups increases during the beginning of the 25-year period studied. After some time, however, the coastal catch surpasses the trawler catch, at approximately the same time as the trawler catch starts to decline. Toward the end of the time period, both vessel groups exhibit decreasing catches due to both discounting and declining stock levels. We see that in the non-cooperative case, both vessel groups exhibit decreasing catches; for all except the first 1±2 years, the non-cooperative catches are well below the optimum.

The effort profiles that land these catches are given in Figure 6.2. We see from this figure that the effort levels peak after 4 and 5 years, respectively, for the trawler and coastal fleets. The average effort of the 5 years around the year that produces the peak number of vessels is 27 trawlers and 274 coastal vessels. These are the more appropriate numbers to use for comparison to the actual amount of effort employed in the fishery. Otherwise, the building-up phase of effort in the beginning and the reduction in effort in the later part of the time horizon of the model is captured. Comparing these numbers to the actual number of vessels employed: 120±200 trawlers and 750±1250 coastal vessels between 1980 and 1994, we see that the model calls for a substantial cut in the number of vessels. Two reasons can be given for this observation. First, even without explicitly taking into account cannibalism, it is generally agreed that the fishery is currently over capitalized. Second, the typical vessel used for the analysis is larger than most of the vessels currently employed in the

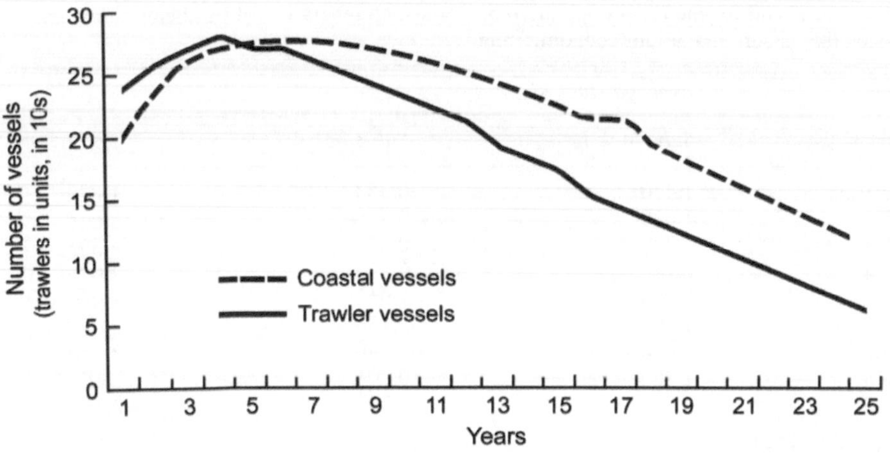

Figure 6.2 Effort profiles for trawler and coastal vessels over a 25-year time period. The coastal vessel numbers are in fleets of 10 vessels.

fishery, and hence has a larger catching capacity than many of the vessels being employed.

Stock sizes

Table 6.4 presents the average sub-stock sizes for different management preferences over the 25-year period. In Table 6.4, we observe that the total stock size that supports the best economic solution occurs when $\beta = 0.6$ at 2.99 million tonnes, with immature and mature sub-stocks of 1.86 and 1.13 million tonnes, respectively. The table also reveals that non-cooperative behavior is disastrous to the health of the stock; total stock size is reduced to a dangerous level of about 0.26 million tonnes, a level that is well below the recommended 0.5 million tonnes minimum spawning biomass for a sustainable cod fishery (Jakobsen, 1993).

In Figure 6.3, the sub-stock profiles over the 25-year period are presented both in the cooperative and non-cooperative cases. We observe in the figure that for the optimal situation, both sub-stocks increase for most of the time period studied. Toward the end of the 25 years, however, the immature sub-stock 1 starts to decline, and is smaller than the mature sub-stock 2 in the last year. There are two possible reasons for the dip in the biomass of sub-stock 2. First, is the fact of discounting the future; second is the fact that the end of the time horizon (or world, if you like) is approaching. Sensitivity analysis using a lower discount rate indicates a less steep decline toward the end of the time horizon. In the non-cooperative case, the sub-stocks decline drastically, with the mature sub-stock being the smallest. At its lowest level, sub-stock 2 is just over 6,000 tonnes.

Table 6.4 Average sub-stock and total stock sizes in million tonnes (over 25 years), for $0 < \beta < 1$, and for the non-cooperative outcomes. Numbers in bold indicate the stock sizes that ensure maximum economic rent

	Average stock sizes		
β	*Sub-stock 1*	*Sub-stock 2*	*Total stock*
0.1	1.470	0.354	1.824
0.2	1.620	0.446	2.066
0.3	1.980	0.615	2.595
0.4	2.140	0.834	2.974
0.5	2.050	1.010	3.060
0.6	1.860	1.130	2.990
0.7	1.680	1.190	2.870
0.8	1.760	0.901	2.661
0.9	1.660	0.707	2.367
Non-cooperative			
	0.178	0.0839	0.2619

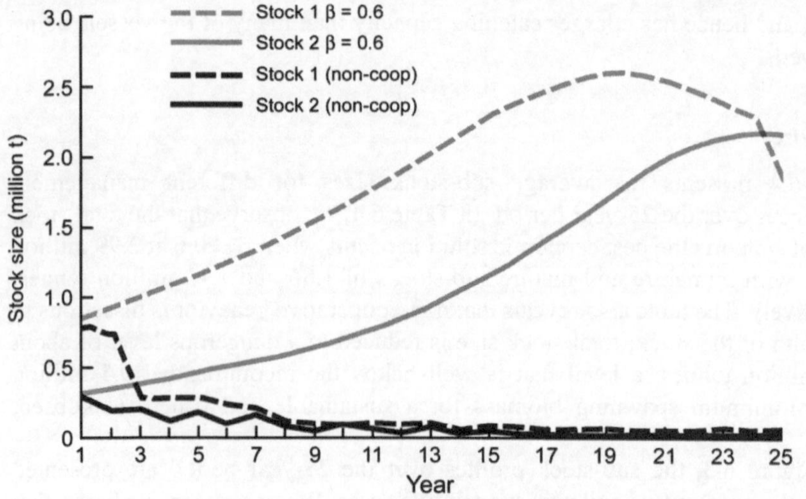

Figure 6.3 Immature and mature sub-stock profiles over a 25-year period for the optimal cooperative and non-cooperative cases. Note that stock 1 and 2 refer to trawl and coastal fleet stock sizes, respectively.

Sensitivity analysis

Table 6.5 presents the results of a sensitivity analysis on costs, prices, growth rates, catchability coefficients, and the discount rate. The sensitivity is measured against profits, average catch, and average sub-stock sizes. Table 6.5 shows that, as expected, with an increase in the cost of fishing, the total profit to the

Table 6.5 Sensitivity analysis: profits, catches, and stock sizes giving maximum economic rent, for an increase in the costs, k_1 and k_2, the prices v_1 and v_2, and the intrinsic growth rates r_1 and r_2, and catchability q_1 and q_2, by 25%, and a reduction in the discount rate, d, from 0.07 to 0.05. The base case in bold defines the optimal results with $\beta = 0.6$. Profits are in billion NOK, while catch and stock sizes are in million tonnes

	Profit			*Average catch*				*Average stock size*		
	Trawl	*Coastal*	*Total*	*Trawl*	*Coastal*	*Total*	*Trawl %*	*Stock 1*	*Stock 2*	*Total*
Base case	**13.35**	**17.36**	**30.71**	**0.198**	**0.252**	**0.450**	**44.0**	**1.860**	**1.130**	**2.990**
$k \uparrow 25\%$	12.13	17.47	29.60	0.190	0.263	0.453	47.9	1.940	1.200	3.140
$v \uparrow 25\%$	17.73	17.38	35.11	0.217	0.211	0.428	50.7	1.600	0.795	2.395
$r \uparrow 25\%$	20.13	30.16	50.29	0.276	0.441	0.717	38.5	2.590	1.460	4.050
$q \uparrow 25\%$	12.91	22.52	35.43	0.188	0.304	0.492	38.2	1.760	1.250	3.010
$d = 0.05$	18.43	20.84	39.27	0.242	0.250	0.492	49.1	1.740	0.858	2.598

two vessel groups declines somewhat. However, the profits to the coastal fleet increase slightly, which means that the overall decline is accounted for by a decrease in trawler profits. Increase in costs also result in increased stock size and a decrease in the trawler fleet share from 44% to almost 42%. These results are presumably due to the larger absolute fishing costs of the trawler vessels (e.g. k_1 is more than ten times as large as k_2). In addition, the reduction in trawler catch, which necessarily follows higher costs, leaves more prey for the mature sub-stock, allowing the coastal vessels to increase their catches while trawler catch share decreases.

Increase in prices results in, as would be expected, opposite effects to those we observe with an increase in costs: a 25% increase in price leads to an increase in overall profits from NOK 30.71 to 35.11 billion. In this case, most of the gains in profits accrue to the trawler fleet. The effects on coastal vessel profits are, however, very small, both in the case of price and cost increases. In the case of price increases, the standing stock size decreases while the catch share to the trawler fleet increases, due to the mature sub-stock declining in size, putting less predatory pressure on the immature sub-stock.

A reduction in the discount rate from 7% to 5% results in an all-round increase in profits, leading to a total increase in resource rent of about 28% from NOK 30.71 to 39.27 billion. Two somewhat surprising observations can be made from Table 6.5 regarding sensitivity to the discount rate. First, the stated decrease in the discount rate leads to about a 5% increase in the trawlers' catch share. Second, as the agents become more patient about the future (i.e. a lower discount rate decreases the value of current versus future revenues, hence making the agents more willing to wait for future revenues), one would have expected the total stock to be allowed to grow larger. However, Table 6.5 shows that this does not appear to be the case. The table also shows that under this scenario, the trawl fleet catch is higher than in the base case scenario, while the coastal vessel catch is lower. A possible reason for the above observations is that this chapter studies the average sub-stock sizes over 25 years. In the case of a decreased discount rate, the agents are less impatient regarding the increase in the sub-stocks, as viewed in Figure 6.3. Hence the sub-stocks increase in size more slowly, but the juvenile sub-stock is, nonetheless, by the end of the 25-year period larger than when the discount rate is higher. This also explains the increased share to the trawler fleet which targets juveniles.

For an increase in the intrinsic growth rates, it was found that profits, catches and stock sizes all increase, as one would expect. It is, however, of interest to note that the trawler catch share is significantly reduced with a change in the growth rates. One reason for this large decrease in trawl catch share is that r_1 (intrinsic growth rate of immature sub-stock) is less than r_2 (recollect that r_1 is a modified intrinsic growth rate). Hence, the growth in the mature sub-stock due to an equal percentage increase in growth rates is relatively greater than the increase in the rate of growth of the immature sub-stock, thereby reducing the optimal catch share of the latter. It appears that the model is quantitatively sensitive to changes in the intrinsic growth rates of the two sub-stocks of cod.

Finally, Table 6.5 reveals that an increase in the catchability coefficients by 25% leads to a decrease in the rent derived from the trawlers, and an increase in the rent from the coastal fleet. The total resource rent increases by over 15%. Similar trends are observed with respect to catch, with the consequence that a significant reduction in the trawler catch share, from 44% to just over 38%, is required. When it comes to sub-stock levels, the table reveals that the immature sub-stock size is lower at 1.76 compared to 1.86 million tonnes in the base case. On the other hand, the mature sub-stock level is higher at 1.25 compared with 1.13 million tonnes in the base case. The model appears to be quantitatively sensitive to changes in the catchability coefficient, but not to the same degree as for changes in the intrinsic growth rates.

To explain the above results, one should note that the catchability coefficient is a measure of the efficiency of the fishing gear, given the availability of the fish targeted. Thus, what these results tell us is that an increase in the efficiency of the coastal fleet by up to 25% will be a welcomed thing, due to the increase in mature sub-stock size and coastal profits. A similar increase in the case of the trawler fleet will be detrimental to both the economics and biology of the fishery, which is what one would expect given the present degrees of efficiency of the two groups of vessels. It seems clear that the model is most robust to changes in the economic parameters, while it is more sensitive to technological and biological parameter changes.

Concluding remarks

Table 6.2 shows that the resource rents derived from the coastal fleet do not increase all the way as β approaches 0, as one would have expected. This is presumably because the low immature sub-stock that emerges for low β values, as a consequence of the nature of the intra-stock interaction in the model, will not give sufficient positive feedback effects on the profits accruing to the coastal group. A larger immature sub-stock (that comes as a result of a larger β) is more to the coastal group's advantage because of the recruitment coefficient a_2. However, this is the case only up to a certain point, where high β values increase the fishing pressure from the trawlers, and also reduce the size of the mature sub-stock.

As stated earlier, the maximum total profit of NOK 30.71 billion is achieved when $\beta = 0.6$. This outcome occurs when the trawler share of the total catch is 44%. Hence, it is economically optimal for the coastal vessels to obtain a greater share of the catch, in order to reduce the predatory pressure on immature cod. The actual trawler catch share of approximately 70% is well above the optimal share computed herein. It is important, however, to keep in mind that in actual fact the trawler and coastal vessels fish, to some degree, upon both sub-stocks. The effect of this on the results is unclear, and is left for future research. With the actual non-Norwegian catch of 55% being taken entirely by trawlers, our results show that not only would it be advantageous for the entire Norwegian quota to be taken by the coastal fleet, but also some of the foreign catch should be taken with passive and active coastal gear, such as nets, hook and line. This puts Norway's somewhat half-hearted efforts at encouraging the development of a Russian coastal fishery in a new perspective.

The actual catch share allocated to the trawlers gives a much lower total stock than the optimum. In our model, a 70% share to the trawlers would result in an average stock size of somewhere between 2.3 and 2.6 million tonnes, while the bioeconomically optimal average stock size is 2.99 million tonnes. Similarly the actual trawler catch share would in equilibrium require a total catch of between 270,000 and 320,000 tonnes, which is well below the optimal size of 450,000 tonnes given by our model. Likewise the profits would be only approximately 60% of the best possible total profits. This clearly demonstrates that the current allocation is sub-optimal. The optimal number of vessels derived from the model is only a fraction of the actual number of vessels in the fishery. The optimal percentage reduction is greater for the trawler group than the coastal vessel group, presumably because the trawlers have historically been given a more than optimal share of the stock.

An important point to note is that the trawlers not only compete with the coastal vessels, but also with the mature sub-stock, as both wish to somehow "catch" the immature sub-stock. Hence, the trawlers obtain the smaller profit when there is no cooperation. Also, as is expected, the profits, catches, and stock sizes are much reduced in the non-cooperative case, compared to the cooperative situation. The actual trawler catch share (approximately 70%) is markedly above the modeled optimal trawler catch share (44%). This is interesting, as the non-cooperative solution is often deemed to be the agents' threat point in a bargaining situation. That is, in a bargaining situation, the non-cooperative solution is the minimum that fishers or agents involved in the bargaining process will accept (Bailey *et al.*, 2010 and the references therein). Even though there is no actual bargaining between trawlers and coastal vessels in the international arena, it is still important to note what pressure the Norwegian coastal fleet could exert on the cod stock if the fishery were to degenerate into a non-cooperative situation. It should be noted, however, that the trawler share in the non-cooperative case is above the share that is allotted to them in the optimal cooperative case: the difference being only about 2.2%. This can be seen as an argument for the trawlers to obtain a larger share than that which our cooperative solution gives.

The bioeconomically optimal total catch is, according to our model, about 450,000 tonnes. This is well below the around 800,000 tonnes that has been caught in recent years. It should be noted, however, that our result also includes low catch levels in the years of the build-up of the stock, and at the terminal periods of the model. This then explains the divergence to the very large catches in recent years. The average annual catch in the 10-year period 1984±1993 is 357,000 tonnes (Anon, 1990a, 1996). In the last few years the catches have risen to about 800,000 tonnes, before declining once again (Anon, 1997.

Comparing the results here with those in Chapter 4, where a Beverton–Holt model is applied, it was found that the non-cannibalistic model gives higher catches and thereby also higher discounted profits than the current model. The lower catches and profits in our case are due to the fact that, all else being equal, cannibalism reduces the stock along with the catches. A corner solution requiring that the coastal vessels in the bioeconomic optimal situation be sole fishers of the resource is obtained in Chapter 4. In the current chapter, corner solutions are not obtained, as it is bioeconomically optimal for both vessel groups to partake in the fishery. The difference between the results from the two studies lies mainly in: (i) the explicit modeling of cannibalism in the current model; and (ii) the different age structure and selectivity patterns assumed in the two models. The results obtained in the present analysis also give far more devastating non-cooperative outcomes due to the same reasons.

The optimal profits and catch shares are especially sensitive to changes in the intrinsic growth rates of the sub-stocks. This underlines the importance of the biological parameters in the model. The trawler catch share is seen to be inversely related to the stock size from the sensitivity analysis on the intrinsic growth rates, with the implication that anything that increases the total average stock size also decreases the trawl catch share. Apparently, this is due to the increased biological predatory pressure upon the immature sub-stock.

The main message from this study is that computational models that explicitly incorporate the cannibalistic tendencies of fish species such as cod need to be developed to help to determine the optimal total allowable catch and its distribution to the different vessel types used to catch the fish. This is especially necessary in fisheries where different vessel types target different sections of the stock. Fisheries biologists in Norway have been skeptical to the use of fishing strategies as a regulatory mechanism for cannibalistic species,[12] stating that cannibalism is a part of the natural regulatory process of a fish population. However, there is little skepticism from the same quarters when it comes to multispecies management, where the interaction between different species is apparently not seen as natural regulation. The results from this study show that it is important biologically and economically to do away with this contradictory stance, whereby interactions within a single species fished by heterogeneous fishers is not seen in a similar light as interactions between species.

7 Implications of implementing an ITQ management system for the Arcto-Norwegian cod stock[1]

Introduction

In theory, at least, most fisheries are managed through the use of biological rather than bioeconomic criteria. Hence, total allowable catches (TACs) are to a greater degree based on maximum sustainable yield (MSY) and not maximum economic yield (MEY). On the other hand, the division of TAC to different fishing groups is usually completely left to political decision-making. Often, however, the allocations to various groups in a fishery have both biological and economic implications, and thus call for the application of bioeconomic criteria for decision-making. A central objective of this work is to quantify the economic loss that may result from sub-optimal determination and allocation of TAC in a fishery. This issue is pursued in the context of the exploitation of the North-east Arctic cod by trawlers and coastal vessels.

In recent years, multispecies issues have received substantial attention from resource economists (see, for instance, Flaaten, 1988; Fischer and Mirman, 1992). This is due to the realization of the complexity of the marine environment, and its effect on the ability of fisheries management to achieve its objectives. Multispecies studies have illustrated the distributional issues involved in the management of fish resources, where the interactions between species are shown to complicate the issues of how much of a species should be fished by which group or fishing gear, in order to enhance the overall economic return from the fishery (Fischer and Mirman, 1992, 1996; Eikeland, 1993). Clearly, the multispecies focus is important, and this chapter illustrates that within species one finds much the same issues that appear between species. For instance, there are interactions within species such as cannibalism which impact the bioeconomics of fisheries in the same manner that predator–prey relations between species do. Also, there are often many different vessel groups targeting different sections of a single stock, in the same way that different fleets catch different interacting species. It is therefore argued that it is only natural that the concern for the impact of multispecies interaction on the bioeconomic outcomes of the fishery be extended to intra-species interactions.[2]

The North-east Arctic cod stock has been shown to have strong intra-stock relations in the form of cannibalism (Eide, 1997). In studying this stock, two things are done. First, the problem of catch allocation is revealed by studying how much of the Norwegian cod quota is landed by the trawlers and coastal vessels.[3] The trawler vessels target mainly the immature section of the stock, which is also preyed upon by the mature section of the stock; hence there is competition between these two "fishers." On the other hand, for the most part only the coastal fleet targets the mature sub-stock. Second, this chapter examines the possible bioeconomic implications of introducing an individual transferable quota (ITQ) management system for the cod fisheries. A key assumption made is that an ITQ management system may lead to a concentration of quotas within one of the two vessel groups. This assumption may seem strong, but as long as there are no barriers to transfers between different groups, where the groups consist of different vessel types and fishing gear, having differing economic viability, concentration of quotas in certain groups may well be expected as a result of the implementation of an ITQ system.[4] Grafton (1996) claims that trade of ITQs appears to be more intense in fisheries where there are significant differences in the gear and vessels used. Geen and Nayar (1988) show that after the implementation of ITQs, purse seine vessels in Australia's southern bluefin tuna fishery increased their share of catch from 16% in 1984 to 42% in 1987. In New Zealand, Dewees (1989) finds that in order to maximize price and minimize cost, 17% of fishers had switched to longlining gear, as a result of ITQs. Furthermore, experience shows that even where a built-in system is put in place to curb the tendency to concentrate quotas in one form or another, quota concentration still seems to result.

This is the case for New Zealand's ITQ system, where it is stipulated that no single company can hold more than 20% of the quota for any species in a management area. Nonetheless, while remaining within the constraints set by this rule, the institution of the ITQ regime led to a concentration of 50% of the quota in the hands of the three largest fishing consortiums in the industry (Cullen, 1996). However, all the above examples of concentration may be bioeconomically optimal. Nonetheless, both social and technological barriers hindering one fisher from owning two vessel types give incentives to stick to the vessel type of first preference, even though this may be sub-optimal.[5] A concentration of quota in one vessel group results in a concentration of catch upon the vessel group's targeted section of the fish stock. This may clearly have negative biological effects, in the sense that a skewed fish stock structure may affect the growth of the stock detrimentally, which again feeds into the economic viability of the fishery.

It is worth mentioning that the implementation of a sustainable and socially acceptable TAC is a crucial guard against overfishing. Economists and fisheries managers expect that the institution of ITQs, based on predetermined TACs, will increase efficiency by reducing the tendency to "over" accumulate fishing capacity, eliminate "the race for fish," reward efficient operators, and improve the ability of fisheries managers to cope with fluctuations in fish stocks and catch rates. There is also hope that a limited number of agents will have less political clout, and/or larger incentives to secure a sustainable fishery. In general and in theory these assertions are valid (Grafton, 1996). However, certain special characteristics of a given fishery may make it less likely for an ITQ system to produce the desired outcomes. Copes (1986) lists 14 issues which he considers to be critical for an ITQ system. Boyce (1992) shows how stock externalities may not be internalized by an ITQ regime. Asche *et al.* (1997) discuss the issue of uncertainty, illustrating how some types of uncertainty result in ITQs not solving the efficiency problem. This chapter focuses on two aspects of a fishery, which to our knowledge have not been studied in connection with ITQs previously, namely, the presence of cannibalism in the cod stock, and the fact that the two vessel types responsible for landing the bulk of cod, target different sections of the stock. Hence there is a technological externality, which is not internalized by an ordinary ITQ system.

This chapter presents the background of the North-east Arctic cod fishery, followed by the development of a bioeconomic model. This is followed by the results of the simulations. Two key results from the analysis are as follows. One, the existing allocation rule (that is, the trawl ladder) applied by Norwegian fisheries managers is not optimal: it functions in a diametrically opposed manner to a bioeconomically optimal allocation rule. This confirms the weaker results in Armstrong (1999), where the focus was solely on the biologically optimal allocation of cod to the two vessel groups. Furthermore, the trawl ladder allocates less of the total catch to the trawler group. Two, if quotas end up being concentrated amongst the coastal vessels in an ITQ system,[6] the mature sub-stock decreases drastically, to well below the biologists' minimum spawning stock requirement of 500,000 tonnes. In the more probable scenario of trawlers obtaining the major part of the quotas, the stock sizes and catch quantities are more stable. Total discounted profits in both ITQ scenarios are, however, well below those theoretically achievable in an otherwise optimally allocated scenario. Finally, a discussion section analyses the results and concludes the chapter.

The North-east Arctic cod fishery

The North-east Arctic cod stock is a migratory fish species that throughout its life span wanders in and out of Russian, Norwegian, and international waters. Hence the catch is also taken by many countries, with Russia and Norway claiming together approximately 90% of the total annual catch. Russia and Norway determine the annual total allowable catch (TAC) and the countries' respective catch shares, cooperatively (Armstrong and Flaaten, 1991a). In recent years, Norway and Russia have divided their common share of the TAC equally. Norway is the only country that employs both coastal and trawl vessels to fish the North-east Arctic cod stock.[7] The other countries apply solely trawl technology. In Norway, the allocation between Norwegian trawlers and coastal vessels is determined through agreements concluded by the Norwegian Fishermen's Association. Here, the trawler and coastal vessel owners come together and determine allocation rules acceptable to them, which are then implemented by Norwegian fisheries managers. The first allocation rule was implemented between 1990 and 1994, and was appropriately called the trawl ladder. The name points to the fact that the shares to the coastal vessel group decreases stepwise as the Norwegian TAC increases. In 1995, a new trawl ladder was agreed upon, which eliminated the steps in the previous agreement.

The new agreement stipulates that, for a Norwegian TAC below 130,000 tonnes the coastal vessel share should be 72%, while for TACs above 330,000 tonnes the coastal vessel share is set at 67%. For TACs between 130,000 and 330,000 tonnes, the coastal vessel share decreases continuously from 72% to 67%. Given that Norway obtains approximately 45% of the TAC, and that all catches other than those by Norwegian coastal vessels are taken using trawl technology, the total trawl share then varies from 67.6% to 69.9%, depending on the size of the Norwegian TAC.[8] This means that the size of the total trawl share is determined by the non-Norwegian share of the TAC (55%) plus the share allotted to the Norwegian trawlers (between 28% and 33% of the remaining 45% of the TAC). It is worth noting that this chapter studies the total trawl shares.

Due to the different fishing technologies applied by the trawlers and coastal vessels, the two face different costs and prices for their landings. Figure 7.1 illustrates the development over time of the two vessel groups' historic weighted costs and profit per tonne. Figure 7.1 shows that the trawler group has lower profits per tonne in each of the years studied, and that for both groups there is an increasing trend in profit per tonne. Despite the per tonne lower cost of trawlers, the fact that they also obtain lower prices per tonne for their catch, results in lower trawler profits per tonne relative to those of the coastal fleet.[9]

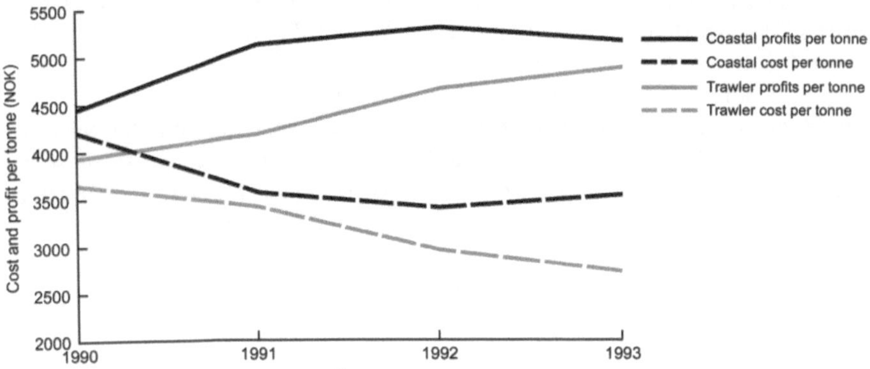

Figure 7.1 Trends in weighted costs and profits in NOK per tonne for Norwegian coastal and trawler fleets, for the years 1990–1993.

Source: Anon., 1990a, 1991, 1992, 1993, 1997.

The model

A deterministic dynamic bioeconomic model is applied, which allows two agents to fish upon separate sections of a given fish stock. The separate sections of the stock interact via cannibalism and recruitment. The analysis concentrates on a single-stock version of the two-stock model, using a structure introduced by Eide (1997). Eide shows that this structure has a close fit to the biological findings regarding the changes in the North-east Arctic cod stock throughout the 1980s. The changes in the biomass levels of the two sections of the stock (hereafter called sub- stocks) are described by the following difference equation:

$$\Delta x_{i,t} = G_i(x_{1,t-1}, x_{2,t-1}) - h_{i,t}, i = 1, 2 \tag{7.1}$$

where $\Delta x_{i,t} = x_{i,t} - x_{i,t-1}$, and $x_{i,t}$, is the biomass of sub-stock i at time t, with i = 1, 2 defining immature and mature sub-stocks, respectively. The rate of catch of sub-stock i is defined as $h_{i,t} = q_i x_{i,t} e_{i,t}$, where q_i is the catchability coefficient of vessel group i, and $e_{i,t}$ is the number of vessels deployed by vessel group i in period t.

The natural growth functions G_i of sub-stocks 1 and 2, which also define the interaction between the two, is described as follows (Eide, 1997):

$$\begin{aligned} G_1(x_{1,t}, x_{2,t}) &= r_1 x_{1,t}\left(1 - \frac{x_{1,t}}{a_1 x_{2,t}}\right) - b x_{1,t} x_{2,t} \\ G_2(x_{1,t}, x_{2,t}) &= r_2 x_{2,t}\left(1 - \frac{x_{2,t}}{a_2 x_{1,t}}\right) \end{aligned} \tag{7.2}$$

The parameters r_i, a_i, and b are all positive constants, where r_i is the intrinsic growth rate of sub-stock i. The parameter a_i describes each sub-stock's recruitment into the other sub-stock, that is, young become old, and old beget young. The parameter b determines the cannibalistic interaction; hence the size of sub-stock 1 is negatively affected by that of sub-stock 2. Since x_j is included in the growth function $G_i (i \neq j)$, as described in the bracketed terms in equation (7.2), there exists a recruitment relationship between immature and mature fish.

It is assumed that the two vessel groups; trawlers and coastal vessels, designated 1 and 2, respectively, fish upon each of their respective sub-stocks, 1 and 2. Thereby vessel group i fishes only upon sub-stock i.[10] The cost function of a given vessel type i in period t, $C(e_i)$, is defined in Chapter 3 as:

$$C(e_{i,t}) = \frac{k_i e_{i,t}^{1+\omega}}{1+\omega} \tag{7.3}$$

where $\omega = 0.01$, and $k_i/(1+\omega) \approx k_i$ is the cost of engaging one fishing fleet (or vessel) for one year. The above formulation introduces concavity in the objective function to help convergence (Flåm 1993) while still maintaining an almost linear cost function.

$$\pi_{i,t} = \pi_i(x_{i,t}, e_{i,t}) = v_i h_{i,t}(x_{i,t}, e_{i,t}) - C(e_{i,t}) \tag{7.4}$$

where v_i is the price per unit weight of sub-stock i. The stream of discounted single period profits of a vessel group, M_i, $i = 1, 2$, can be defined as:

$$M_i(x_i, e_i) = \sum_{t=1}^{T} \delta^t \pi_i(x_{i,t}, e_{i,t}) \tag{7.5}$$

where δ is the discount factor. Note that $t = 1, \ldots, T$ represents fishing periods, with T denoting the last period. Given a cooperative regime, the goal of the agents is to find a sequence of effort, e_i, and sub-stock levels, $x_{i,t}$, $i = 1, 2$, such that a weighted average of their respective objective function is maximized:

$$\pi(x_1, e_1, e_2) = \beta M_1(x_1, e_1) + (1 - \beta) M_2(x_2, e_2) \tag{7.6}$$

subject to equation (7.2) above, and the obvious non-negativity constraints. β and $(1 - \beta)$ denote the preferences of the two players, that is, it is an indication of the weight that is put on the objective functions of 1 and 2, respectively. A β greater than 0.5 gives more weight to the management preferences of agent 1, over those of agent 1 (Munro 1979). To compute the optimal path of effort, the solution procedure applied is described in Flåm (1993) and applied herein.

Data

The cost parameter k_i, denoting the cost of engaging the most cost effective vessel in each group for one year, is calculated to be NOK[11] 21 and 1.53 million for trawlers and coastal vessels, respectively. Following Armstrong (1999), the catchability coefficients q_i, $i = 1, 2$, are determined using the number of vessels, the actual catch, and the sub-stock size given by the model. This method gives catchability coefficients of 0.00117 and 0.00665 for $i = 1, 2$, respectively. Prices for fish used in the computations are the average ex-vessel prices for cod delivered by the trawlers and coastal vessels in 1992, that is, 7,579 and 8,655 NOK per tonne, respectively.

The biological data are taken from Eide (1997), which gives the intrinsic growth rate, r, for the immature and the mature sub-stock to be 0.5003 and 0.6728, respectively. Furthermore, the parameter a is given values of 8.7608 and 1.1880, for immature and mature sub-stocks, respectively, while b is set at 0.2023. The 1990 immature and mature stock sizes, determined to be 783,900 and 280,500 tonnes, respectively, are used as the initial stock sizes in the model.

Results

The optimal versus the trawl ladder allocation results of the maximization of the weighted profit function in equation (7.6) are presented. That is, given the manager's choice of weighting of the two vessel groups' preferences (i.e. profits), the optimal catch and effort paths, and catch shares for each value of β are determined.[12]

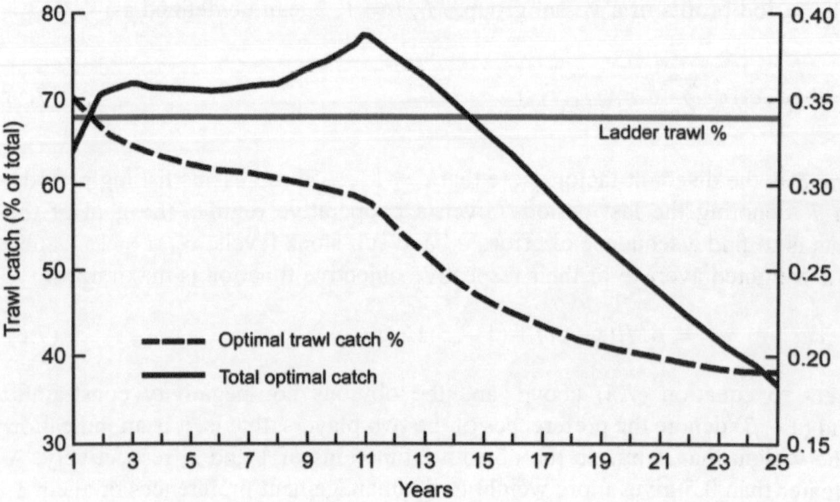

Figure 7.2 The computed optimal TAC (when $\beta = 0.7$), the computed optimal catch share, and the trawl ladder catch share to the trawl fleet over time.

Figure 7.2 illustrates the overall optimal catch and total trawler share over a 25-year time period. The β value that determines this overall optimal catch is 0.7, as this gives the highest discounted profit. Hence, this implies that optimal management should put more weight on the management preferences of the trawler group than on those of the coastal vessel group. Figure 7.2 shows that for $\beta = 0.7$, optimal trawler shares decline throughout the time period, while the trawl ladder shares resulting from the same level of the TAC as in the optimal case, lead to relatively constant, higher catch shares throughout the time period. It is, however, important to remember that the trawl ladder shares are here applied to the optimal TAC (emerging for $\beta = 0.7$), which would *not* emerge if catch shares were as prescribed by the trawl ladder. This point is made clearer in the figures that follow.

Discounting appears to be an important factor explaining the sharp declines seen in Figure 7.2. Sensitivity analysis shows that for a decreased discount rate, the total catch does not decline as quickly as in the above. Also, the trawler catch share does not drop as sharply. Another point to note is that as the game approaches the end of the time horizon, players accelerate their depletion of the stock since the implication to them is that the "end of the world" is approaching. It is therefore more appropriate to compare the trawl ladder catch with the average catches from the model in the early part of the game. For $\beta = 0.9$, the average allocation of the TAC to the trawlers of approximately 74% was obtained, which is the closest to the *actual* trawl ladder percentage allocation. Hence, it appears that in practice, fisheries managers weigh trawler preferences substantially higher than coastal vessel preferences.[13]

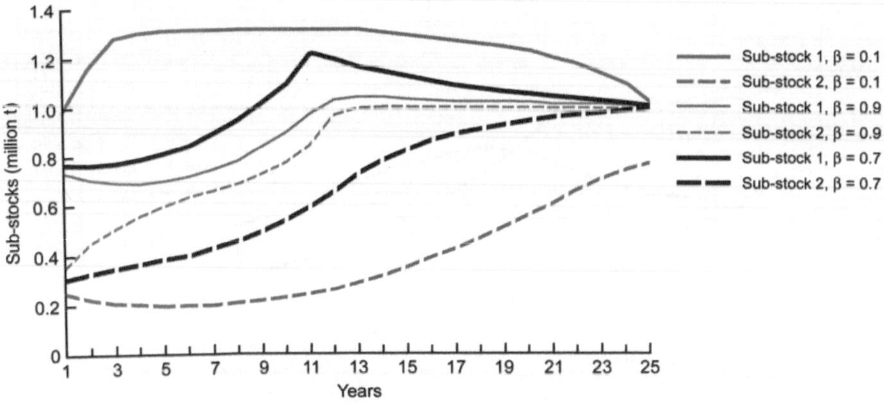

Figure 7.3 Sub-stock sizes over time for $\beta = 0.1$ (the coastal vessels buy up the ITQs) and $\beta = 0.9$ (the trawlers buy up the ITQs), and the optimal case ($\beta = 0.7$) (sub-stock 1 is immature, while sub-stock 2 is mature).

An ITQ management regime

To mimic the situation in an ITQ management regime, β is set at 0.1 and 0.9,[14] respectively; the former 3 value depicting a concentration of the TAC in the coastal vessel group, while the latter concentrates the quota in the trawler group. The resulting sub-stock sizes are depicted in Figure 7.3, which shows that in the case of the coastal vessels obtaining most of the quota ($\beta = 0.1$), the mature sub-stock is reduced dramatically to levels well below the biologists requirement of a minimum spawning stock of 500,000 tonnes. The immature sub-stock size is at its highest level in this case.

As shown in Figure 7.4, the total catch given that the coastal vessels obtain most of the quota, is well below the alternative when the trawlers obtain the largest part of the quota, for most of the time period.

Table 7.1 shows that, irrespective of which vessel group buys up the quota rights, there is an overall economic loss under an ITQ regime. As far as the coastal vessel group is concerned, the optimal outcome with $\beta = 0.7$ is the most preferable. Somewhat surprisingly, the next best outcome for them occurs under a quota regime when $\beta = 0.9$. This gives the trawlers most of the catch. The trawl group prefers, marginally, a quota regime in which it holds most of the quota ($\beta = 0.9$). The second best outcome for this group is achieved when the optimal allocation ($p = 0.7$) is implemented.

A comparison of the total net present value of NOK 21.33 billion (obtained over a 25-year period when $\beta = 0.7$) with NOK 16.45 billion (obtained when $\beta = 0.9$, approximation of the trawl ladder), shows a computed economic loss of almost NOK 5 billion.

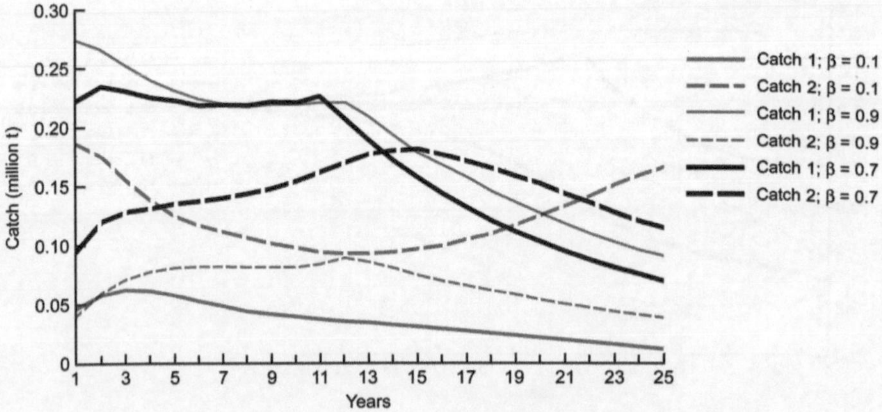

Figure 7.4 Catch of sub-stocks 1 and 2 over time, for $\beta = 0.1$ (the coastal vessels buy up the ITQs) and $\beta = 0.9$ (the trawlers buy up the ITQs), and the optimal case ($\beta = 0.7$) (sub-stock 1 is immature, while sub-stock 2 is mature).

Table 7.1 Profits in billion NOK (present value over 25 years), for ($\beta = 0.1$, 0.7, and 0.9) (β refers to the preferences of the trawl fleet)

	Profit		
β	*Trawl*	*Coastal*	*Total*
0.1	2.64	4.20	6.84
0.7	10.73	10.60	21.33
0.9	10.77	5.69	16.45

Discussion

The results presented above show that the existing allocation of the North-east Arctic cod between trawlers and coastal vessels leads to some economic losses. Compared to the optimal case, the trawl ladder allocation rule produces shares that are opposite to the bioeconomically optimal allocation. In addition, an approximation to the actual allocation gives a profit reduction of more than NOK 5 billion over a 25-year period, compared to what is achievable when the optimal weighting of the two vessel groups is implemented. These results are partly due to the fact that the two vessel groups are modeled to target their respective sub-stocks, which interact naturally through cannibalism. For an increasing stock size, the mature sub-stock increases its predatory activity upon the immature sub-stock, making it optimal to increase the catch of the mature sub-stock, thereby increasing the coastal vessels' catch share.

It was found that when two vessel groups fish upon separate sub-stocks, which interact cannibalistically, the implementation of an ITQ regime may not be bioeconomically optimal. That is, if the ITQ regime results in a concentration of the quota within one vessel group, the biological advantage of fishing with both vessel types is lost. The study shows, however, that it is relatively better economically for the quotas to be concentrated within the trawl group in an ITQ management system, as this gives the highest total profits over time.

Furthermore, this chapter shows that the allocation rule currently in place is to the trawler's advantage. If this favorable weighting of the trawl preferences is a deliberate management policy, then the trawl ladder allocation rule produces about the best outcome management can expect, with the gap in resource rent between this and the optimal taken as the price management is willing to pay for having such a preference. If, however, such a deliberate policy is non-existent, then the trawl ladder puts too much weight on the trawler's preferences, in which case appropriate management action is needed. It is, however, not surprising that somewhat greater weight is put on the preferences of trawlers considering that Russia operates only trawlers, and Norway has a substantial fleet. Finally, it is important to note that since this model is computational, it mainly gives insights based on the information available at the time. Re-runs of the model will be necessary from time to time to account for important changes that may occur, for instance, to markets and vessel group structure, which may result from the introduction of ITQs in the fishery (Wilen and Homans, 1997). The development of management models that solve the technological externality in an ITQ system, as illustrated here, is a part of the author's plans for future work.

8 Marine protected area performance in a game-theoretic model of the fishery[1]

Introduction

Marine protected areas (MPAs) are parts of the marine habitat in which fishing is controlled or prohibited entirely for all or part of the time (Bohnsack, 1990 Sumaila *et al.*, 2000; Sumaila and Charles, 2002). The interest in MPAs as a tool for fisheries and ecosystem management has now gone past marine researchers and conservation groups to policy makers. Evidence of this is the May 2000 Executive Order issued by the President of the USA calling for "appropriate actions to enhance or expand protection of existing MPAs and establish or recommend, as appropriate, new MPAs."[2] In fact, a few countries (USA and Australia) have recently declared large MPAs.

Among the groundwork recommended to guide how to go about implementing the Executive Order is the "assessment of the economic effects of the preferred management solution."[3] The objective of this chapter is precisely to provide an assessment of the economic performance of MPAs: will the establishment of an MPA bring about significant biological and economic benefits if the management objective is to maximize the joint profits of fishers? What sizes of MPAs may be considered optimal when the fishery is managed jointly and separately?

Published economic models that study the potential economic benefits of MPAs can be grouped into (i) single-species/non-spatial/single agent (sole owner) models (e.g. Holland and Brazee, 1996; Hannesson, 1998; Sumaila, 1998b); (ii) single-species/spatial/single agent models, (e.g. Holland, 1998; Sanchirico and Wilen, 1999); (iii) multispecies or ecosystem/spatial/single-agent models (e.g. Walters, 2000; Pitcher *et al.*, 2000); and (iv) multispecies or ecosystem/non-spatial/single-agent models (e.g. Sumaila, 1998b). To our knowledge, there are no multi-agent models that explore the economic potential of MPAs in the literature. The current chapter fills this gap by developing a two-agent model for the assessment of MPA performance. With a two-agent model, an important question is addressed, which until now has not been addressed in the literature, namely, how will MPAs perform when participants in a fishery cooperate, resulting in efficient management, versus when they do not cooperate, leading to competitive and wasteful management.

The North-east Atlantic cod fishery is used to demonstrate the workings of the model developed. This cod stock is highly migratory, working its way through both Norwegian and Russian Exclusive Economic Zones (EEZs), as well as international waters. Norway and Russia together determine the total allowable catch (TAC), giving each country approximately 45% of the TAC, with the remainder fished by other countries, such as Iceland, the Faroe Islands, and some EU countries. The Russian and other-country catch is mainly fished by trawlers offshore, while the Norwegian share of the TAC is divided between two vessel groups – trawlers and coastal vessels. Thus, the fishery is presently managed cooperatively (Nakken *et al.*, 1996) which makes the current model relevant for studying the fishery.

The model

Biological aspects

Let recruitment of age 0 fish to the whole habitat in period t ($t = 1, \dots, T$), R_t, be represented by the following Beverton–Holt recruitment function.[4]

$$R_t(B_{t-1}) = \frac{\alpha B_{t-1}}{1 + \gamma B_{t-1}} \tag{8.1}$$

where $B_{t-1} = \sum_{a=1}^{A} p_a w_{s,a} n_{a,t-1}$ represents the post-catch spawning biomass of fish; p_a is the proportion of mature fish of age a ($a = 1, \dots, A$); $w_{s,a}$ is the weight at spawning of fish of age a; $n_{a,t-1}$ is the post-catch number of age a fish in period $t-1$; and α and γ are constant biological parameters. The α and γ values determine the recruitment for a given spawning biomass, which again determines the pristine stock level.

Initially, it is assumed that the stock and recruits are homogeneously distributed and randomly dispersed at a constant density. The fish population is split into two distinct components, $i = 1, 2$ where 1 and 2 denote the protected and unprotected areas, respectively. There is net movement from the protected to the unprotected area, due to fish density being high relative to the carrying capacity in the protected section of the habitat (the Basin model; MacCall, 1990). This movement is captured by the net migration rate, which tells us the net proportion of a given age group of fish that is transferred from the protected to the unprotected area in a given fishing period.

The division of the habitat is done by first dividing the initial stock size between the protected and unprotected areas in proportion to these areas' respective sizes. Hence, an MPA consisting of 20% of the habitat results in a split of the initial stock size into a 2:8 ratio in favor of the unprotected area. Second, it is assumed that recruitment takes place separately in the two areas defined in equation (8.1) above, each area with its own spawning biomass $B_{i,t-1}$ and γ_i, $i = 1, 2$. The α parameter, being an intrinsic element of the stock under consideration, is kept equal for fish both in the reserve and in the fished area. Finally, the respective γ parameters are set such that (i) the sum of recruitment from both areas satisfies

$$R_t^1 + R_t^2 = R_t \quad \text{for} \quad B_{t-1}^1 + B_{t-1}^2 = B_{t-1} \tag{8.2}$$

and (ii) the recruitment into the protected and unprotected areas is directly related to the quantity of the total biomass in them. These conditions are enforced by giving γ^i values from 1 to 10, depending on the MPA size, with a value of 1 depicting a large MPA and a value of 10 depicting a small MPA.

For the protected area, the stock dynamics in numbers, $n_{a,t}^1$, is described by

$$\begin{aligned} n_{0,t}^1 &= R_t^1 \\ n_{a,t}^1 + \psi n_{a,t}^1 &= s n_{a-1,t-1}^1, \quad \text{for} \quad 0 < a < A \\ n_{A,t}^1 + \psi n_{A,t}^1 &= s(n_{A-1,t-1}^1 + n_{A,t-1}^1), \quad n_{a,0}^1 \ \text{given} \end{aligned} \tag{8.3}$$

where A is the last age group of cod, the parameter s is the age independent natural survival probability of cod; $\psi n_{a,t}^1$, is the net migration of age a cod from the protected to the unprotected area in period t, and ψ is the net migration rate; $n_{a,0}^1$, denotes the initial number of age a cod in the protected area. Recollect that there is no fishing in the protected area.

The stock dynamics in the unprotected area are expressed as

$$\begin{aligned} n_{0,t}^2 &= R_t^2 \\ n_{a,t}^2 + h_{a,t}^2 &= s n_{a-1,t-1}^2 + \psi n_{a,t}^1, \quad \text{for} \quad 0 < a < A \\ n_{A,t}^2 + h_{A,t}^2 &= s(n_{A-1,t-1}^2 + n_{A,t-1}^2) + \psi n_{A,t-1}^1, \quad n_{a,0}^2 \ \text{give} \end{aligned} \tag{8.4}$$

where $h_{a,t}^2$ is the total catch function, defined in the traditional way as

$$h_{a,t}^2 = q_a n_{a,t}^2 e_t$$

where q_a is the age dependent catchability coefficient and e_t is the effort employed in the fishing of cod in period t.

I introduce a shock in the natural system (Sumaila, 1998b) by incorporating a recruitment failure (zero recruitment) that occurs in each of the years 5 to 15 of the 28-year time horizon model. That is, recruitment failure occurs in each year within this range of years. It is important to note that the shock is assumed to occur only in the fished area, an assumption which follows Lauck (1996), where it is assumed that true uncertainty occurs due to human intervention in the natural environment, leading to over-fishing and habitat degradation. Sensitivity analysis is performed to study the effects of changes in the degree of shock and the exchange rate.

Economic aspects

A dynamic model is applied to describe the joint and separate management of the North-east Atlantic cod fishery in which there are two participants, namely, the coastal vessel group (C) and the trawler gear group (T). These are the two main vessel types used to catch cod. The single period profit from fishing, $\Pi_m(.)$, is defined as

$$\Pi_m(n^2, e) = v\sum_{a=0}^{A} w_a q_a n^2_{a,t} e_t - \frac{k_1}{1+\omega}(e_t)^{1+\omega} \tag{8.5}$$

where m = C, T.[5] The variable e_t (t = 1, 2, ..., T = 28) denotes the profile of effort levels employed by the particular player; n^2 is the age and time dependent stock size matrix in the fished area; v is the price per unit weight of cod; w_a is the average weight of age a cod; k is a cost parameter, and $\omega > 0$ is a parameter introduced to ensure strict concavity in the model, which is required to ensure convergence (Flåm, 1993).

I assume that under joint management, the objective of the participants in the fishery is to find a sequence of total effort levels, e_t (t = 1, 2, ..., T = 28), that would maximize their joint benefits. Using the effort level as the control variable, the vessel groups jointly maximize their present value of profit,

$$\text{prof}_j = \sum_{t=1}^{T} \delta_j^t \Pi_{j,t} \tag{8.6}$$

where $\delta = (1+r)^{-1}$ is the discount factor and r denotes the discount rate. The optimization is carried out for given sizes of the MPA, subject to equations (8.2), (8.3), and (8.4), and the obvious non-negativity constraints.

Under separate management, I assume each agent wishes to maximize his or her own profits, that is, Π_C and Π_T, respectively, for the coastal and trawler fleets. The non-cooperating agents must therefore choose their own effort levels in each fishing period in order to maximize own discounted profit, given that the other agent does the same. This is done without regard to the consequences of their own actions on the other agent's payoff. For the coastal fleet this translates into choosing their own effort level to maximize

$$\text{prof}_c = \sum_{t=1}^{T} \delta_c^t \Pi_{c,t} \tag{8.7}$$

Modified Lagrangian functions, in the sense of Flåm (1993), are set up and computed using the simulation package Powersim (Byrknes, 1996). The computational procedure is resorted to because it is difficult to solve the current multi-cohort model analytically (Conrad and Clark, 1987).

The solution procedure (algorithm) is from non-smooth convex optimization, in particular, subgradient projection and proximal-point procedures (see for example, Flåm, 1993). This class of algorithms is intuitive because they are of "behavioristic" type: they model out-of equilibrium behavior as a "gradient" system driven by natural incentives.

The data

The parameters α and γ are set equal to 3 and 1 per billion kilograms, respectively, to give a billion age zero fish (assuming negligible weight at age zero) when the spawning biomass is half a million tonnes.[6] Based on the reported survival rate of cod (Nakken *et al.*, 1996), s is given a value of 0.81 for all a. The price, v is equal to NOK 6.78 and 7.46 per kilogram of cod landed by trawlers and coastal vessels, respectively. The cost parameter, k_m, which denotes the cost of engaging a fleet of vessels (10 and 150, respectively, for T and C) for one year, is calculated to be NOK 210 and 230 million, respectively, and ω is set equal to 0.01. The discount factor is given a value of 0.935, as recommended by Norway Bank.

The initial number of cod age groups 1 to 8 is obtained by taking the average of the initial numbers from 1984 to 1991 (reported in Table 3.12 of the ICES, 1992). For the other age groups, I assume the same number as for age group 8. This gives the initial numbers report in Table 8.1. The parameter $p_a = 0$ for $a < 7$ and 1, otherwise. See Table 8.1 for data on catchability coefficients, weight in catch and other parameters of the model.

Results

Plots of the resource rent and standing biomass as a function of the MPA size are presented in Figure 8.1 for both the joint and the separate management scenarios. The figure shows that the total resource rent from the fishery is strongly related to the size of the MPA. The rent increases with the MPA size until an optimal size is reached at 60% and 70% under separate and joint management, respectively. With regard to standing biomass, a similar pattern is observed: total standing biomass, in both the protected and fished areas, increases with increasing MPA size. But contrary to what one would have expected, it peaks at the same MPA sizes as in the case of the resource rent. One would have expected the standing biomass to keep increasing linearly with size but this is not the case. The reason for this counterintuitive result is that after 60% and 70% of the habitat has been protected under separate and joint management, respectively, optimal fishing in the unprotected area requires a much lower standing biomass in this part of the habitat, which is low enough to more than compensate for the higher biomass in the protected area.

The base case results for key outputs and decision variables of the model (discounted profits, standing stock biomass and MPA size) are presented in Table 8.2. The table reports the outcomes for "with" and "without" an MPA under both separate and joint management. In the case of the "with" MPA case, the MPA size that gives the highest discounted profits is reported.

Table 8.1 Parameter values used in the model

Age a (years)	*Catchability coefficient q(p, a)*		*Weight in catch (a)*[a] *(kg)*	*Initial numbers*[b] *(millions)*
	T	*C*		
1	0	0	0.10	460
2	0	0	0.3	337
3	0	0	0.6	298
4	0.032	0	1	223
5	0.062	0	1.4	117
6	0.075	0	1.83	61
7	0.084	0.056	2.26	33
8	0.084	0.140	3.27	9
9	0.084	0.191	4.27	9
10	0.084	0.255	5.78	9
11	0.084	0.217	7.96	9
12	0.084	0.153	9.79	9
13	0.084	0.089	11.53	9
14	0.084	0.051	13.84	9
15	0.084	0.0255	15.24	9

The parameter p_a is given the value (0, 0, 0, 0, 0.02, 0.06, 0.25, 0.61, 0.81, 0.93, 0.98, 1, 1, 1, 1) for $a = \{0, 1, \ldots, 15\}$.

a $w(s,a)$ is assumed to be 90% of the weight in catch.

b These are obtained by taking average initial numbers of various age groups from 1984 to 1991 reported in Table 3.12 (ICES, 1992).

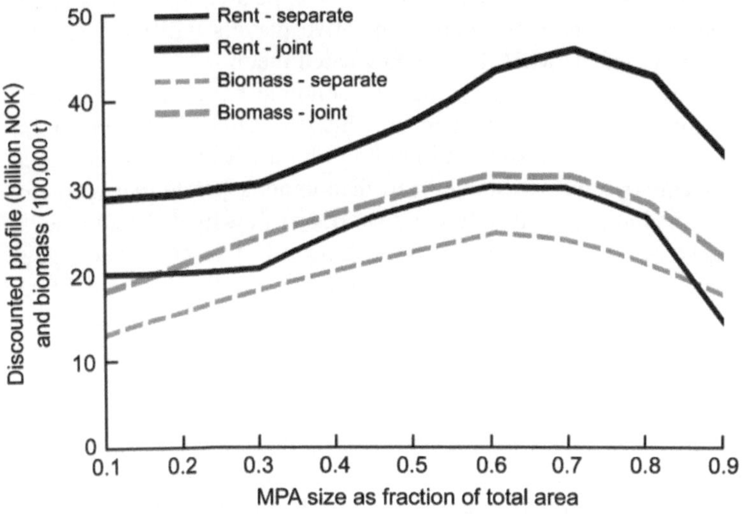

Figure 8.1 Rent and standing biomass as a function of MPA size.

Table 8.2 Base case: total discounted profits (in billion NOK), the average annual standing biomass (in million tonnes) and MPA size in percentage of habitat area, and discount factor of 0.935

		Non-cooperative	*Cooperative*
Discounted profits (no MPA)	Trawlers	13.93	18.15
	Coastal	12.60	16.82
	Total	26.53	34.97
Discounted profits (best MPA)	Trawlers	13.77	23.70
	Coastal	16.50	22.37
	Total	30.27	46.06
Average stock biomass	No MPA	1.15	1.81
	Best MPA	2.48	3.16
	MPA size (%)	60	70

Under the assumptions of the model, MPAs are likely to give higher discounted profits in a fishery that is likely to face a shock. Under non-cooperative management fishers make a total of about NOK 30.27 billion with an MPA, compared with NOK 26.53 billion without an MPA. This is achieved with an MPA size of 60% of the habitat. The equivalent numbers under joint management are NOK 46.06 and 34.97 billion, respectively. In this case, the optimal MPA size is 70%. To reveal the insurance value of MPAs under the two management regimes, these numbers were compared to the discounted profits that would be obtained when the habitat is assumed not to face a shock. This comparison showed that MPAs manage to protect about 62% of the no shock returns to the fishery under joint management, and 68% under separate management. This suggests that MPAs have an insurance value, which appears to be greater under separate management: At least, the non-cooperative players will have to wait until the fish come out of the MPA before they catch them.

It should be noted that, in general, higher economic benefits are achieved under joint than under separate management. This is because fishers in a joint management setting allow the resources to build to higher levels after the shock has occurred, by employing less fishing effort than under separate management, especially during the initial periods of the time horizon of the model (Figure 8.2). On average, between 28% and 35% more fishing effort is employed under separate than under joint management. More fish are left in the sea "with" than "without" an MPA (Table 8.2). Hence, the implementation of MPAs can protect and enhance the stock biomass by helping maintain high standing fish biomass under the scenarios explored.

More fish are left in the sea under the joint management regime because fishers here already have an efficient management policy in place; hence, they are in a better position to reap benefits from the insurance cover that MPAs provide. This result leads to two interesting observations. First, fisheries with good management plans can, under certain situations, benefit from implementing MPAs. Second, MPAs are no panacea; they need to be implemented as complements to other traditional management tools.

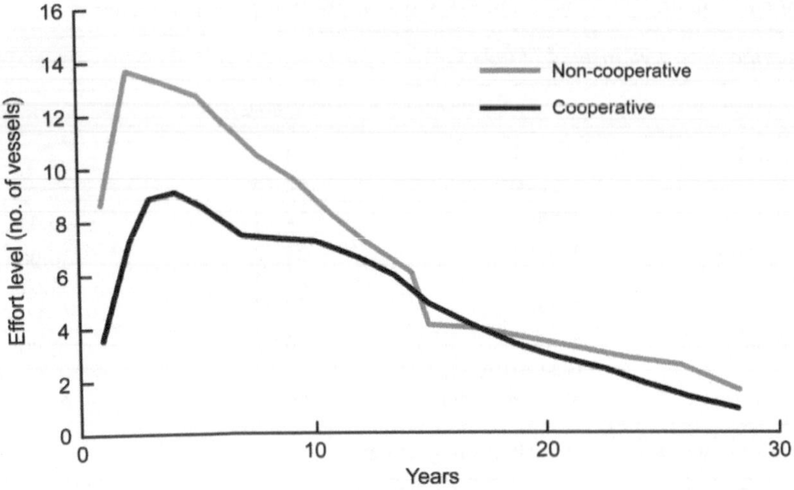

Figure 8.2 Effort profile under cooperative and non-cooperative management.

The discount factor, the exchange rate between the protected and unprotected areas, and the degree of shock introduced in the model were varied to examine how sensitive the model results are to changes in these parameters. The optimal MPA sizes remain the same except when a milder recruitment failure is assumed, and only under separate management (Table 8.3). In which case, the optimal MPA size changes from 60% to 50%. An interesting result from the sensitivity analysis is that at a low discount rate (2%), MPAs do not appear to enhance economic benefits. This is an indication that MPAs are more likely to protect economic benefits only when discount rates are high. Hence, MPAs may be a means by which to mitigate the negative effects of high discount rates in fisheries. This means that when fishers are very impatient, e.g. in developing countries because of the pressures of meeting basic needs, or when a fishery is operating under open access, MPAs could be employed as a tool to protect the stock, and mitigate economic waste.

Conclusion

The current analysis, as non-definitive as it may be because it is computational, suggests that MPAs can help reduce losses in resource rent for a fishery in a real world situation, where shocks to the habitat are bound to happen from time to time. The establishment of MPAs could help maintain high fish biomass in the marine habitat. This is the case whether fishers behave cooperatively or not. Hence, this study brings to the fore the insurance value of MPAs, as argued by, among others, Clark (1996) and Lauck (1996).

Based on the specifics of the model presented, it appears that for the full economic benefits of reserves to be realized, they have to be implemented as

Table 8.3 Sensitivity analysis: total discounted profits (in billion NOK), average annual standing biomass (in million tonnes) and MPA size as percentage of habitat area

The discount factor is 0.98 instead of 0.935			
		Separate	*Joint*
Discounted profits (no MPA)	Trawlers	23.57	48.91
	Coastal	29.18	54.52
	Total	52.74	103.41
Discounted profits (best MPA)	Trawlers	24.32	36.00
	Coastal	25.84	41.61
	Total	50.17	77.60
Average biomass	No MPA	0.91	2.50
	Best MPA	2.12	2.91
	MPA size (%)	60	70
Lower migration rate of 0.4 versus 0.8 in protected areas			
Discounted profits (no MPA)	Trawlers	13.93	18.15
	Coastal	12.60	16.82
	Total	26.53	34.97
Discounted profits (best MPA)	Trawlers	11.90	17.80
	Coastal	12.79	16.46
	Total	24.69	34.26
Average biomass	No MPA	1.15	1.81
	Best MPA	2.79	3.43
	MPA size (%)	60	70
Milder shock – recruitment failure from year 5 to 9			
Discounted profits (no MPA)	Trawlers	13.07	17.05
	Coastal	11.30	15.28
	Total	24.37	32.33
Discounted profits (best MPA)	Trawlers	14.77	24.09
	Coastal	16.10	22.32
	Total	30.87	46.41
Average biomass	No MPA	1.36	2.02
	Best MPA	2.50	3.17
	MPA size (%)	50	70

part of an efficient management package. The chapter isolates the differences in economic and biological outcomes, depending on whether the fishery is managed jointly or separately. Finally, this chapter shows, again based on the specifics of the model, that MPAs could serve as useful fisheries management tools when fishers have high discount rates, and are therefore very impatient.

9 Distributional and efficiency effects of marine protected areas[1]

Introduction

This chapter studies the efficiency and distributional effects of implementing marine protected areas (MPAs) in the North-east Atlantic cod fishery in the Barents Sea. Recent work on MPAs has focused on the biological and economic efficiency of implementing this form of management (Polacheck, 1990; Sanchirico and Wilen, 1999). However, when MPAs are introduced in areas where there exists extensive and varied use of the marine resources, then clearly this management measure may have distributional effects, which could cause resistance to its implementation. Bohnsack (1993) argues that marine reserves reduce conflict between user groups via physical separation of fishery and non-fishery interests. It is shown that the implementation of a marine reserves map cause conflict *within* diverse fishery interests. In the biological enthusiasm and perhaps more critical economic focus upon MPAs, issues regarding distributional effects have been afforded little attention. Holland (2000) shows some distributional effects clue to dislocation. It is demonstrated that even without actually forcing some agents to move their activity, the age structure of a stock and the selectivity patterns of the catch when an MPA is implemented may result in changes in the payoffs to different fishing groups.

Studies have shown that under certain conditions less catch is obtained with the implementation of MPAs (e.g. Hannesson, 1998). Other works have also indicated that catch may in fact increase with MPAs compared to a no-MPA scenario in a patchy system (Sanchirico and Wilen, 2001) or if the marine ecosystem is likely to face a sudden shock due to true uncertainty (Lauck *et al.*, 1998; Sumaila, 1998a). However, it is conceivable that there may exist winning and/or losing groups of fishers with the implementation of MPAs, something demonstrated to be the case. Furthermore, it is shown that some management systems combined with MPAs may ensure that the more efficient vessels gain increased access to the stock, resulting in an increase in the total benefits from the resource. In the case of fisheries with complex agent and stock interactions, the implementation of an MPA may have *implicit* distributional effects where, for instance, a quota management system is more *explicit*.

In this chapter, physical location is ignored, and focus is rather placed on how the size of an MPA may affect different vessel groups through their gear selectivity patterns. In the case of increases in catches resulting from the implementation of an MPA, these may not be distributed in an acceptably equitable fashion amongst the agents in a fishery. In a situation where an MPA results in a loss in total benefits, it may nonetheless be found that some groups enjoy increases in their own private benefits. In some cases there may be clear winners and losers, or just big and small winners (or losers). In many cases, the different agents in the fishery will be aware of how they may be affected by the implementation of a marine reserve, resulting in substantial lobbying against such a management regime. In other cases, the resulting effects may be more uncertain, and the inequity of the regime may not appear prior to the regime being instituted. The costs and political tensions in both situations are apparent.

This chapter undertakes a case study of a single species fishery, namely, the North-east Atlantic cod fishery, which involves mainly two different fisher groups – the trawler and the coastal vessel fishers. Several studies of MPAs have discussed the issue of what management regime should exist outside the MPA, and the importance of this to the expected outcome. Focus has been on open access (e.g. Pezzey *et al.*, 2000; Hannesson, 1998; Holland and Brazee, 1996) rather than some form of management (see however, Reithe, 2002 for models in which other private property management regimes have been incorporated). The presence of open access or unmanaged common property outside the boundaries of the MPA will naturally not maximize the resource rents.

As an illustration of a non-managed fishery, the non-cooperative case is applied,[2] while the cooperative case illustrates a managed fishery. It is worth noting that cooperative management is often seen in the context of international fisheries management. However, the same concept has been applied to the management of domestic fisheries with different fisher groups. Furthermore, a shock to the biological system is included in the model. Sumaila (1998b) illustrates how an MPA can function as a hedge against shocks or natural fluctuations in a stock. The shock is presented as a deterministic recruitment failure over a 10-year period, which may introduce different distributional effects for vessels targeting different age groups within a stock.[3]

Key results from our analysis include (1) depending on the *ex ante* status quo and *ex post* management a win-win, lose-lose, or win-lose situations may emerge with the implementation of an MPA; and (2) the two vessel groups may not prefer the same MPA sizes.

In the next section, the North-east Atlantic cod fishery is presented, followed by the model and data used in the analysis, and the results obtained. A sensitivity analysis is undertaken to ascertain the robustness of the results with respect to key biological and economic parameters of the model. The chapter concludes with a discussion of the results.

The North-east Atlantic cod fishery

The North-east Atlantic cod fishery has a long history of boom and bust. Though never as dramatic as fisheries for small pelagic fishes, the variation in the cod stock size has, over the last 50 years, been substantial; from 4 to 0.3 million tonnes (Anon., 2000a). Effective regulation of the trawler fleet has been in place since the early 1970s, while the coastal vessel fleet has only been effectively managed since the late 1980s. Nonetheless, a large stock variation is observed. The last serious decline was in 1989, when the fish biomass was reduced to less than 20% of its estimated highest levels. Hence, despite the fact that total allowable catches (TACs) and actual catches have not been substantially over the biologist's recommendations, the cod stock still seems to vary to a certain degree (Anon., 2000a). The reasons given for this variation have been illegal fishing and bycatch, cannibalism and predation, as well as atrophic changes in the environment. In this scenario of uncertainty, MPAs are one of the possible management tools.[4]

The North-east Atlantic cod stock is a highly migratory fish stock, spending periods of its life cycle in Norwegian, Russian, and international waters. The cod is caught by both Russia and Norway, as well as a group of other nations mostly allotted quotas by these two countries. Fishing is carried out using a wide variety of different vessel types and gears. It is, however, common to divide these vessels into two distinct groups, namely trawlers and coastal vessels, which is the vessel group division applied in this study. The trawler vessel group is a relatively homogeneous entity, while the coastal vessel group consists of a large diversity of vessel sizes and gear types, which is aggregated in the model for tractability. To a certain degree, the two vessel groups target different age groups within the cod stock, as a large part of the trawler catch is concentrated on the younger cod, while the coastal vessel catches mainly consist of mature cod (Armstrong, 1999). This is due to both gear selectivity and the migration pattern of the cod stock; as the young grow up in the open sea, migrating in to the coast to spawn. The difference in catch configuration as well as different degrees of freshness and processing onboard, leads to markedly different prices for the catches of the two vessel groups.

The model

Allow recruitment of age 0 fish to the whole habitat in period t ($t = 1 \ldots T$), R_t, to be represented by the following Beverton–Holt recruitment function.[5]

$$R_t(B_{t-1}) = \frac{\alpha B_{t-1}}{1 + \gamma B_{t-1}} \qquad (9.1)$$

where $B_{t-1} = \sum_{a=1}^{A} p_a w_{sa} n_{a,t-1}$ represents the post-catch spawning biomass of fish; p_a is the proportion of mature fish of age a ($a = 1, \ldots, A$); w_{sa} is the weight at spawning of fish of age a; $n_{a,t-1}$ is the post-catch number of age a fish in period $t-1$; and α and γ are constant biological parameters. The α and γ values determine the recruitment for a given spawning biomass, which again determines the pristine stock level.

Initially it is assumed that the stock and recruits are homogeneously distributed, and randomly dispersed at a constant density.[6] The fish population is split into two distinct components, $i = 1, 2$, where 1 and 2 denote the protected and unprotected areas, respectively. It is assumed that there is net movement from the protected to the unprotected area, due to build-up of biomass in the protected section of the habitat, or more generally due to sink-source relationships between the two areas. Alternatively, improved habitat conditions due to reduced catch pressures could explain this effect (Rodwell *et al.*, 2003). This movement is captured by the *net migration rate*, ψ which, assumed to be constant here, captures the net proportion of a given age group of fish that is transferred from the protected to the unprotected area in a given fishing period.[7]

The division of the habitat is done by, first, dividing the initial stock size between the protected and unprotected areas in proportion to these areas' respective sizes. Hence, an MPA consisting of 30% of the habitat, results in a split of the initial stock size into a 3:7 ratio in favor of the unprotected area. Secondly, it is assumed that recruitment takes place separately in the two areas defined as in equation (9.1) above, each area with its own B^i_{t-1} and γ^i, $i = 1, 2$. The α parameter, being an intrinsic element of the stock under consideration, is kept equal for fish both in the reserve and in the fished area. Finally, the respective γ parameters are set such that (1) the sum of recruitment from both areas satisfy

$$R^1_t + R^2_t = R_t \quad \text{for} \quad B^1_{t-1} + B^2_{t-1} = B_{t-1} \tag{9.2}$$

and (2) the recruitment into the protected and unprotected areas is directly related to the quantity of the total biomass in them. These conditions are enforced by giving γ^i values from 1 to 10, depending on the MPA size, with a value of 1 depicting a large MPA and a value of 10 depicting a small MPA.

For the protected area, the stock dynamics in numbers, $n^1_{a,t}$, is described by

$$\begin{aligned} n^1_{0,t} &= R^1_t \\ n^1_{a,t} &= s n^1_{a-1,t-1} - \psi n^1_{a,t}, \quad \text{for } 0 < a < A \\ n^1_{A,t} + \psi n^1_{A,t} &= s(n^1_{A-1,t-1} + n^1_{A,t-1}), n^1_{a,0} \text{ given} \end{aligned} \tag{9.3}$$

where the parameter s is the age independent natural survival probability of cod; $\psi n^1_{a,t}$ is the net migration of age a (where A is the last age group) cod from the protected to the unprotected area in period t, and ψ is the net migration rate; $n^1_{a,0}$ denotes the initial number of age a cod in the protected area. Recollect that there is no fishing in the protected area.

The stock dynamics in the unprotected (fished) area, $n^2_{a,t}$, is expressed as

$$n^2_{0,t} = R^2_t$$
$$n^2_{a,t} = sn^2_{a-1,t-1} + \psi n^1_{a,t} - h_{a,t} \quad \text{for } 0 < a < A$$
$$n^2_{A,t} + h_{A,t} = s(n^2_{A-1,t-1} + n^2_{A,t-1}) + \psi n^1_{A,t}, n^2_{a,0} \text{ given} \tag{9.4}$$

where $h_{a,t}$ is the total catch function, defined in the traditional way as

$$h_{a,t} = q_a n^2_{a,t} e_t$$

where q_a is the age dependent catchability coefficient, e_t is the effort employed in the fishing of cod in period t. One of the central justifications for implementing MPAs is hedging against uncertainty (Clark, 1996). In order to illustrate the effect of uncertainty upon the distributional effects of an MPA, a sudden shock is introduced in the natural system (Sumaila, 1998b) by incorporating a recruitment failure (zero recruitment) that occurs in each of the years 5 to 15 of the 28-year-time horizon model. It is important to note that the shock is assumed to occur only in the fished area, an assumption which follows Lauck *et al.* (1998), where it is argued that true uncertainty occurs due to human intervention in the natural environment, leading to overfishing and habitat degradation, which again can have effects such as described in our model. Sensitivity analysis is performed to determine the effects of these assumptions on the results of our analysis.

Economic aspects

A dynamic game theoretic model is applied to describe the cooperative and non-cooperative management of the North-east Atlantic cod fishery in which there are two participants, namely, the coastal vessel group (C) and the trawler gear group (T). These are the two main vessel types used to catch cod. The single period profit from fishing, $\Pi_m(.)$ is defined as

$$\Pi_m(n^2, e) = v \sum_{a=0}^{A} w_a q_a n^2_{a,t} e_t - \frac{k_1}{1+\omega}(e_t)^{1+\omega} \tag{9.5}$$

where m = C, T (C stands for coastal fleet, and T is the trawler fleet).[8] The variable e_t (t = 1,2, …, T = 28) denotes the profile of effort levels employed by the particular player; n^2 is the age and time dependent stock size matrix in the fished area; v is the price per unit weight of cod; w_a is the average weight of age a cod; k is a cost parameter, and $\omega > 0$ is a parameter introduced to ensure strict concavity in the model, which is required to ensure convergence (Flåm, 1993).

It is assumed that under cooperation, the objective of the participants in the fishery is to find a sequence of total effort levels, e (t = 1, 2, …, T = 28) that maximizes their weighted joint discounted resource rent from the resource for given MPA sizes, as a function of the net migration rate. Thus, using the effort level as the control variable, the vessel groups jointly maximize their present value of profit, prof

$$\text{prof} = \sum_{t=1}^{T} \delta^t (\beta \Pi_{cf} + (1-\beta)\Pi_{tf}) \quad (9.6)$$

where $\delta = (1+r)^{-1}$ is the discount factor, β is the weighting of the preferences of the coastal fleet and $(1-\beta)$ is that of the trawler vessels, with respect to the use of the fish (Munro, 1979), and $r$denotes the discount rate. Hence $\beta = 1$ describes the management preferred by the coastal vessels winning the day, while the trawler preferences are not taken into account, and vice versa with $\beta = 0$. The optimization is carried out for different sizes of the MPA, subject to equations (9.2), (9.3), and (9.4), and the obvious non-negativity constraints.

It should be noted that β is only a bargaining parameter and not a decision variable in our model. β is exogenous to the model, but it is varied in our simulations in order to determine the *optimum optimorum,* that is, the maximum joint present value profit, as described in equation (9.6). Also worth noting is the fact that cooperative solutions can be with side payments in which a player can be bought out of the fishery if such an action will lead to an increase in the overall payoff, and those without side payments, where buying-out a player is not permitted. Results under a cooperative management scenario with, and without side payments, with the former producing the *optimum optimorum* (Munro, 1979) are presented.

It is important to stress at this juncture that for a cooperative outcome to be willingly implemented by the players in the game, they have to meet the individual rationality principle. That is, it must be the case that each player does better by cooperating. In other words, the players must receive higher payoffs under cooperation, than they would receive under their threat point equilibrium solutions. In some cases to achieve cooperation, side payments will have to be paid to one party in the game. Similarly, under non-cooperation it is assumed that each agent wishes to maximize own profits, that is Π_C and Π_T, respectively, for the coastal and trawler fleets. The non-cooperating agents must therefore choose their own effort levels in each fishing period in order to maximize own discounted profit, taking into account the reactions of the opponent in the game (Levhari and Mirman, 1980; Fischer and Mirman, 1992). This is done without taking into account the consequences of their own actions on the other agent's payoff. For the coastal fleet this translates into choosing own effort level to maximize

$$\text{prof}_c = \sum_{t=1}^{T} \delta_c^t \Pi_{c,t} \quad (9.7)$$

The solution to the non-cooperative model serves as the threat point for the cooperative solutions presented earlier in this chapter.

A computational procedure to solve the model was applied because it is generally difficult to solve a multi-cohort model such as the current one analytically (Conrad and Clark, 1987).

The solution procedure (algorithm) is based on non-smooth convex optimization, in particular, subgradient projection and proximal-point procedures (Flåm, 1993). These classes of algorithm are intuitive because they are of a "behavioristic" type; they model out-of-equilibrium behavior as a gradient system driven by quite natural incentives, in this case marginal profits. This algorithm is described in detail in the Appendix.

Data

The parameters α and γ are set equal to 3 and 1 per billion kilograms, respectively, to give a billion age zero fish (assuming negligible weight at age zero) when the spawning biomass is half a million tonnes.[9] Based on the survival rate of cod, s is given a value of 0.81 for all a. The price is $v =$ NOK 6.78 and 7.46 per kilogram of cod landed by trawlers and coastal vessels, respectively. The cost parameter k_m, which denotes the cost of engaging a fleet of vessels for one year is calculated to be NOK 210 and 230 million, respectively, and ω is set equal to 0.01. For the sake of scaling, units of fishing effort of 10 trawlers and 150 coastal vessels are used to calculate the (fleet) cost parameters for the two vessel groups (Kjelby, 1993).

The discount factor is given a value of 0.935. The initial numbers of cod in each age group, the maturity parameters, the catchability coefficient, weight at catch and spawning are given in Table 9.1. The net migration rate ψ is set to be 0.8. A sensitivity analysis is carried out for the discount rate, the recruitment failure, and the net migration rate to determine how they may affect the results of the study.

Table 9.1 Total market values (discounted profits) in billion NOK totaled over the 28-year simulation period, average annual standing biomass in million tonnes, and MPA size as a percentage of habitat

		Non-cooperative	*Cooperative*
Discounted profits (no MPA)	Trawlers	13.93	18.15
	Coastal	12.60	16.82
	Total	26.53	34.97
Discounted profits (best MPA)	Trawlers	13.77	23.70
	Coastal	16.50	22.37
	Total	30.27	46.06
Average stock biomass	No MPA	1.15	1.81
	Best MPA	2.48	3.16
	MPA size (%)	60	70

Notes: Migration rate = 0.8.
Discount factor = 0.935.

Table 9.2 Change in discounted profits depending on *ex ante* or *ex post* management

	Ex post *management with best MPAs*	
Ex ante *management with no MPAs*	*Non-cooperation*	*Cooperation*
Non-cooperation	T ↓ C ↑	T ↑ C ↑
Cooperation	T ↓ C ↓	T ↑ C ↑

Notes: T, trawlers.
C, coastal vessel group.

The results

An overview of the results of the analysis relating to stock size and the distribution of discounted profit to the two vessel groups for the different scenarios is presented in Table 9.1. When the vessel groups operate under cooperative management, higher discounted profits and higher average stock biomass emerge compared to the non-cooperative scenario. Table 9.2 illustrates the different distributional effects that may result from the implementation of an optimal MPA, depending on the management regime originally in place and the management regime applied to the now reduced fishable area with an MPA. The results show that win-win, win-lose and lose-lose outcomes may emerge depending on whether there exists cooperation or not.

Economic results

From Table 9.2, it is observed that when the *ex post* management is cooperative with an MPA, both vessel groups receive more discounted profits, regardless of prior management without an MPA. As in Sumaila (1998b), the total discounted profit increases when an MPA is established in the presence of a shock, but what is new is that this study also shows that all agents involved gain from the MPA. This is, however, not necessarily the case when the *ex post* management is non-cooperative. In this case, only the coastal vessels gain from the introduction of an MPA, and this only when the *ex ante* management is also non-cooperative. Furthermore, both vessel groups lose when an MPA is introduced, if the *ex post* management is non-cooperative, while the *ex ante* management was cooperative. On the quantitative side, the coastal vessel group enjoys the higher gain, both under cooperation and non-cooperation.

Figures 9.1 and 9.2 present the discounted profits of the two vessel groups for a spectrum of MPA sizes. From the two figures, it is seen that cooperation outside the MPA results in a higher optimal MPA size than without cooperation. In the cooperative case, both vessel groups prefer an MPA size of 70%. In the non-cooperative case, the trawlers prefer an MPA size of 50%, while the coastal vessels prefer 80%.

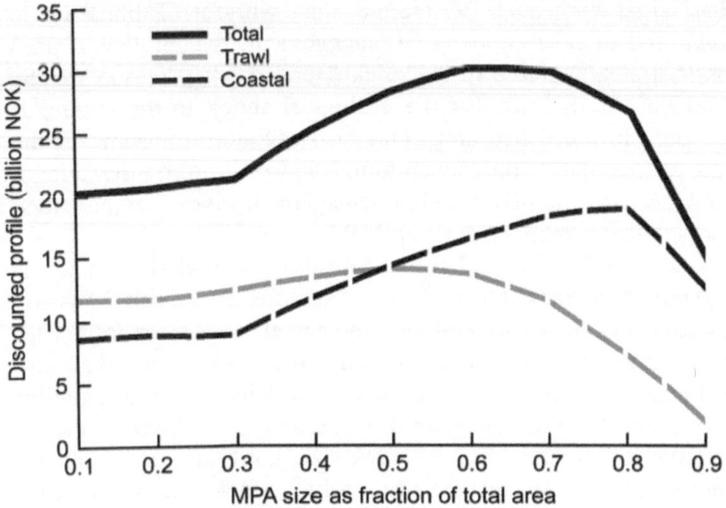

Figure 9.1 Discounted profits to trawlers and coastal vessels for different MPA sizes, in the case of non-cooperation.

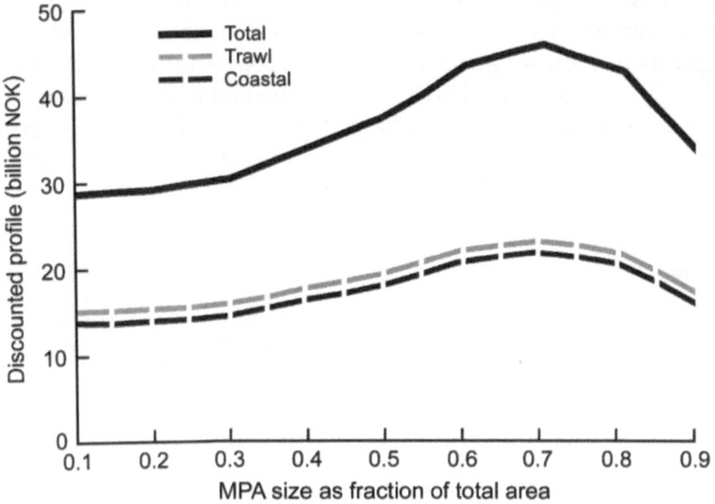

Figure 9.2 Discounted profits to trawlers and coastal vessels for different MPA sizes, in the case of cooperation.

Biological results

The stock size when management is both *ex ante* and *ex post* non-cooperative is more than 50% larger with an MPA regime, than without (Table 9.1). In the case of *ex ante* and *ex post* cooperative management, the implementation of an MPA leads to a 40% increase in the stock. Here, the key reason for the higher standing biomass is the fact that the ecological shock to the system is assumed to occur only in the fished area. This also accounts for some of the conservation gains we see under non-cooperation. It is worth noting that a large enough MPA combined with non-cooperative management gives a higher stock size than cooperative management without an MPA.

The stock sizes described in Table 9.1 are close to the actual stock sizes of the North-east Atlantic cod stock during different periods in the last 50 years. The MPA stock sizes in Table 9.1 fit well with the actual stock sizes from after the Second World War up until the mid-1970s, when the stock averaged around 3 million tonnes. Fishing was limited during the war, creating conditions similar to a marine reserve. From the 1970s onward, stock sizes are closer to the no MPA non-cooperative management regime stock size, until strict management regimes came into place in 1989, making the resulting stock size resembles a cooperative outcome. Following a build-up phase, stocks reached levels close to, but mostly below the no MPA cooperative level presented in Table 9.1.

Effort levels

The total average effort employed with an MPA is lower than without an MPA, regardless of management choice (Table 9.3). Furthermore, the fishing effort under cooperation is lower than that under non-cooperative management. However, the effort level used when there is cooperation and no MPA, exceeds the effort without cooperation and with a large MPA. Note also that, in the cooperative case, it is optimal to have no trawl effort. That is, it is optimal to buy out the trawlers.

Sensitivity analysis

Sensitivity analysis is carried out in order to determine the percentage change in the results when the discount factor is increased to 0.98; the net migration

Table 9.3 Average effort use (over a 28-year period) in number of vessels

	No MPA		*Best MPA*	
	Coastal	*Trawler*	*Coastal*	*Trawler*
Non-cooperation	651	36	479	23
Cooperation	924	0	740	0

rate ψ is decreased to 0.4, and recruitment failure is assumed to occur only in years 5–9.

Table 9.4 shows that these relatively large changes in the parameters do not affect the optimal MPA sizes, except when the degree of shock is reduced. In this case, however, the non-cooperative MPA size is reduced from 60% to 50%. The sensitivity analysis shows that effects upon the average stock biomass are also limited. Nonetheless, when the discount factor increases the size of the discounted profit changes, as is to be expected.

Discussion

The results reported herein indicate that the implementation of MPAs may result in varying distributional effects, depending on the management regime in place before and after the MPA is implemented. In a situation with a shock to the system, and where there is cooperative interaction between the agents after the implementation of an MPA, it is seen that *both* vessel groups gain from the change. However, if an MPA is introduced in combination with non-cooperation, this may not ensure gains to all agents involved. The results show furthermore that the advantageous character of MPAs (in economic terms) in the presence of shocks or true uncertainty may be diminished when the management outside the MPA is non-cooperative.

The impact of MPAs on the average biomass is significant. This is especially so in the case of non-cooperation, where the biomass is increased by more than 50% when an (optimal) MPA consisting of 60% of the total habitat is introduced. The reason for this is that in the non-cooperative situation without an MPA, the stock is heavily fished down. Hence, even a small MPA increases the biological wellbeing of the stock substantially. In light of the discussion around open access and MPAs (Hannesson, 1998), it is worth noting that an MPA with non-cooperative management gives a higher stock size than purely cooperative management without an MPA. Hence, despite sub-optimal management outside the marine reserve, the protection of the stock is higher than under optimal management without a reserve. This illustrates the biological attractiveness of MPAs.

With the possibility of shocks to the system, there are winners and losers relative to their optimal possibilities. For instance, under non-cooperation, the coastal vessels lose out the most when the optimal MPA size of 60% is chosen, compared to their optimal choice of MPA size. The overall trend shows that the trawlers would prefer smaller MPAs than the coastal vessels. This is due to the two vessels' catch strategies, where the trawlers catch immature cod, which if left in an MPA yield mature cod, which again can migrate and be caught by the coastal vessels. This would not be the case if, for instance, a breeding ground such as the Lofoten Islands in Norway was closed to fishing. It appears that the assumption of homogeneous distribution is a prerequisite for the above result.

Comparing our results to the actual catch of North-east Atlantic cod, existing management is closer to the non-MPA cooperative management results without

Table 9.4 Sensitivity analysis: percentage change in the results when the discount factor, δ, is increased to 0.98, the net migration rate, ψ, is decreased to 0.4, and the recruitment failure is reduced to years 5–9

		$\delta \uparrow$ *5%*		$\psi \downarrow$ *50%*		*Shock* $\downarrow$ *50%*	
		Non-cooperation	*Cooperation*	*Non-cooperation*	*Cooperation*	*Non-cooperation*	*Cooperation*
Discounted	Trawlers	69	169	–	–	−6	17
profits	Coastal	132	224	–	–	−10	16
(no MPA)	Total	99	196	–	–	−8	17
Discounted	Trawlers	77	52	−14	−25	7	44
profits	Coastal	57	86	−22	−26	−2	26
(best MPA)	Total	66	68	−18	−26	2	35
Average	No MPA	−21	38	–	–	18	48
stock _biomass	Best MPA	−15	−8	13	10	1	22
	MPA size (%)	0	0	0	0	−17	14

side payments. That is, the trawlers are not bought out. Hence, the only overall acceptable management option, within the options studied, would be a cooperative MPA, as this increases both the trawler and the coastal vessel present value profits. The optimal MPA size without side-payments is however unchanged at 70%, and the optimal stock sizes are almost identical. Catch is shared such that the trawlers obtain 55% of the catch, with the coastal vessels catching the remainder. However, the real world catch share to the coastal vessels is lower than what this analysis prescribes, as the actual trawler versus coastal vessel catch ratio is approximately 7:3.[10] Hence, a move to an optimal MPA with only 5.5% of the total catch in the cooperative setting would presumably be a hard pill for the actual trawlers of today to swallow.

The fact that it is optimal to have zero trawl catching in the cooperative case (i.e. buying out the trawlers with side-payments) is most probably because complex biological interactions such as cannibalism are not incorporated in the model. When cannibalism is introduced to this fishery, it is not optimal to allow only one vessel group to catch. Hence, more complex intra-stock relations than used in this model may show that it is optimal for both fleet groups to exploit the resource, rather than to effectively create a reserve of the whole offshore area.

The fact that both vessel groups prefer cooperation with a 10% MPA to non-cooperation without an MPA is an interesting observation. This indicates the potential of even a small MPA if there is *ex ante* non-cooperation. Even without cooperation outside the MPA, both groups would prefer a 50% reserve to the threat point described by non-cooperation without a reserve. This illustrates the potential of marine reserves in poorly managed fisheries. Even when management outside the reserve is hard to implement, a reserve may well be preferred to a badly managed non-reserve fishery. However, the North-east Atlantic cod fishery is highly regulated, hence such a possible improvement in the fishery is hard to imagine. The result does, however, present some hope for externally unmanaged fisheries, though this could naturally be stated to be the case with any form of alternative effective management. It is also of interest to note that the discounted total profit in the cooperative situation without an MPA does not exceed the discounted total profits in the non-cooperative situation with an MPA by much.

The resulting optimal MPA sizes show a degree of robustness to substantial changes in key parameter values. However, as would be expected, changes in the discount factor have significant effects on the magnitude of the discounted profits and the standing biomass, while reductions in the net migration parameter and recruitment failure have only limited effects. Furthermore, the distributional effects – that is, who wins and who loses – vary somewhat with changes in parameter values. The political problems surrounding MPA implementation become quite clear in this context. From these results, it appears that the claim that the implementation of MPAs will resolve conflicts amongst fishers (Bohnsack, 1993) is not well founded. On the contrary, the implementation of MPAs may lead to conflicts due to the widely divergent distributional effects.

Let us conclude with some general observations: when a marine habitat is likely to face shocks (which seems to be the case in reality), ill participants in the fishery may well benefit from the establishment of an MPA, depending upon the status quo management in place. However, there is also the possibility that only some of them will benefit depending on the management regimes in place both before and after MPA creation. The desired MPA size for the agents involved may also differ. These results are derived from a model in which agents specifically target either mature or immature fish. Similar distributional issues as described in this chapter may be expected to arise in situations where agents fish a single cannibalistic species, or fish in a competitive or prey-predator multispecies system.

10 Playing sequential games with Western Central Pacific tuna stocks[1]

Introduction

The marine life in the Western Central Pacific Ocean (WCPO) contains four main tuna stocks – albacore, skipjack, bigeye, and yellowfin. Skipjack and albacore tuna fetch the lowest price among the four tuna stocks and are usually processed into cans for lower-end markets. On the other hand, bigeye and yellowfin tuna are caught for the high-end sashimi market in Japan, where they command high prices per tonne. Even though there are Taiwanese and Korean purse seine fleets active in the WCPO, purse seines are used mainly by domestic countries (including the Philippines and Indonesia) in the region, and primarily target skipjack, but also capture significant amounts of juvenile bigeye and yellowfin tuna as bycatch. In selling their catch, the owners of the purse seine fleet do not distinguish between the two more valuable tuna stocks from skipjack – they sell them all at the same much lower skipjack price per tonne. The purpose of this chapter is to study the economic effects of this interaction between the two fleets in a game theoretic modeling framework.

We develop a model that captures this relationship; thereafter, we employ a numerical method to compute equilibrium solutions from the model. First, Nash non-cooperative equilibrium solutions are determined when the fleet types, longlines and purse seines, are assumed to be managed separately by their respective owners (the current situation in the WCPO tuna fisheries). Second, we identify joint management equilibrium solutions by assuming that management of the two fleets are carried out by a joint management body. The latter solution is best in the sense that the "sole owner" is expected to internalize the externalities that are bound to originate from taking juvenile bigeye and yellowfin tuna as bycatch by the purse seine fleet. The main reason for the high incidence of juvenile bigeye and yellowfin is that the purse seine fleet uses fish aggregating devices (FADs) to help them reduce their cost of fishing while increasing their catching efficiency, which is beneficial to them but quite detrimental to the overall potential benefits from the yellowfin and bigeye stocks.

Specifically, the main questions with which we are concerned with in this chapter are: (i) What is the maximum discounted economic rent that can be sustainably derived from the tuna resources of the WCPO under joint and separate management? (ii) How significant is the difference between these two solutions?

(iii) What is the effect of exploitation on the stocks of skipjack, bigeye, and yellowfin under these management regimes? (iv) How are catches of the three species traded-off against each other, given changes in prices, costs, and discount factors?

The application of game theory to study fisheries accelerated after the publication of Munro (1979). Many applications have since been published in the literature, from theoretical (Levhari and Mirman, 1980; Flaaten, 1988) to empirical works (Clark and Kirkwood, 1979, Sumaila, 1995). Some papers have focused on ecological externalities (Fischer and Mirman, 1992), others on gear interactions and externalities (Sumaila, 1997b), a few on market externalities (Dockner *et al.*, 1989), and several on dynamic externalities (e.g. Dudley and Waugh, 1980; Kaitala and Pohjola, 1988). Recent developments include the application of characteristic function games (Kronbak and Lindroos, 2007) and partition function games (Pintassilgo, 2003). Few sequential fishing games exist in the literature (e.g. Hannesson, 1995; Laukkanen, 2003), and to our knowledge none of them focus on gear interactions in terms of juvenile bycatch of valuable tuna stocks as addressed in this contribution. Similarly, there have been papers in the literature that have modeled tuna fisheries using game theory (e.g. Kennedy and Watkins, 1986) but we are not aware of any sequential game theoretic models focusing on the impact of using FADs on juvenile bycatch of highly priced tuna stocks.

The model

We model the WCPO ecosystem focusing on the three fleets (purse seine, shallow water longline, and deeper water longline), targeting three main tuna stocks – skipjack, bigeye, and yellowfin – supported by the ecosystem. The source of externality in our model is the fact that purse seiners used to target skipjack also catch significant quantities of juvenile bigeye and yellowfin, mainly because they use FADs to help reduce their cost of fishing. We first describe the stock dynamics of the three tuna stocks and then we build a sequential game theoretic model on this.

Stock dynamics

Let the set $a = \{0, \ldots, A\}$ be age group of tuna stock $i = \{\text{skipjack, yellowfin, bigeye}\}$, where A is the last age group, which is different for each i; and the set $t = \{1, \ldots T\}$ denotes the set of fishing periods, where T is the terminal period, set equal to 25 years here, allowing us to make predictions for the next quarter of a century. For a given year class of skipjack, the number of individuals decreases over time due to natural and fishing mortalities, hence, we have

$$N_{0,t} = f(B_{i,t-1}),$$

$$N_{a,i,t} + h_{a,i,t} = s\, N_{a-1,i,t-1}, \quad \text{for} \quad 0 < a < A; t > 0,$$

$$i = \text{ skipjack, yellowfin, bigeye}$$

$$N_{A,i,t} + h_{A,i,t} = s\,(N_{A,i,t-1} + N_{A-1,i,t-1}), \tag{10.1}$$

where $f\left(B_{i,t-1}\right)=\frac{\chi B_{i,t-1}}{1+\gamma B_{i,t-1}}$ is the Beverton–Holt recruitment function for tuna stock i[2]; $B_{i,t-1}=\sum_{a} p_a w_{a,i,t-1} n_{a,i,t-1}$ represents spawning biomass in weight; $p_{a,i}$ is the proportion of mature fish of age a of tuna species i; χ and γ are constant biological parameters[3]; $s_{a,i}$ is the constant age-specific survival rate of tuna, and $n_{a,i,t}$ represents the post-catch number of age group a tuna in fishing period t of tuna species i. The function, $h_{a,i,t}$, is the catch of age a of tuna species i in fishing period t.

Catch functions

Let p = {PS, LLS, LLD} denote the players in the game, where PS, LLS and LLD refer to the owners of purse seines, shallow longlines, and deepwater longlines, respectively. Then the catch of age group a skipjack in fishing period t,is straightforward and given by

$$h_{a,i,t}=q_{a,i}n_{a,i,t}e_t^{\mathrm{PS}} \quad \text{where} \quad i=\text{skipjack} \tag{10.2}$$

Here, e_t^{PS} is the fishing effort exerted on skipjack using purse seines (PS); and $q_{a,i}$ stands for the age dependent catchability coefficient of the purse seine fleet.

The catch of age group a bigeye in fishing period t, is a bit more complicated because the purse seine fleet takes them as bycatch. The catch is therefore expressed as follows:

$$h_{a,i,t}=q_{a,i}n_{a,i,t}e_t^{\mathrm{LLD}}+\alpha_a(\mathrm{FAD},\, n_{a,i,t},\, e^{\mathrm{PS}}) \quad \text{where} \quad i=\text{bigeye} \tag{10.3}$$

Here, e_t^{LLD} is the fishing effort exerted on bigeye using the deep water longline fleet (LLD); and $q_{a,i}$ stands for the age dependent catchability coefficient of this fleet. The parameter α_a denotes the bycatch of age a bigeye by the purse seine fleet, which is assumed here to depend on the number of FADs used and the stock size of bigeye tuna.

The catch of age group a yellowfin in fishing period t, is even more complicated because not only do the purse seine fleet take them as bycatch, they are also targeted by both longline fleets. The catch is therefore expressed as follows:

$$h_{a,i,t}=q_{a,i}n_{a,i,t}e_t^{\mathrm{LLS}}+q_{a,i}n_{a,i,t}e_t^{\mathrm{LLD}}+\theta_{\mathrm{a}}(\mathrm{FAD},\, n_{a,i,t}) \quad \text{where} \quad i=\text{yellowfin} \tag{10.4}$$

Here, e_t^{LLS} and e_t^{LLD} are the fishing efforts exerted on yellowfin using the shallow (LLS) and deep water longline fleets (LLD), respectively; and $q_{a,i}$ stands for the age dependent catchability coefficient of these fleets. This parameter plays a central role in this model: it is the device used to account for the special features of the different fisheries. A procedure for estimating this parameter is given in Sumaila (1997). The parameter θ_a is the bycatch of age a yellowfin by the purse seine fleet, which is assumed here to depend on the number of FADs used and the stock size of yellowfin tuna.

Price and cost of fishing

The demand for tuna is assumed to be perfectly elastic, thus the stock-dependent price per kilogram of fish, v_i, which denotes either the price of skipjack, bigeye, or yellowfin, is assumed to be constant. Following Sumaila (1995), the cost of fishing by a given player i in period t, $C(i,t)$, is modeled as an "almost" linear function of its fishing effort (number of fleets), $e_{i,t}$:

$$C_{p,t}, = \frac{k_p e_{p,t}^{1+w}}{1+w} \tag{10.5}$$

where $\omega = 0.01$, and $k_{i/(1+b)} \approx k_i$ is the cost of engaging one fishing fleet defined in the data section) for one year by purse seines and longlines. This formulation of the cost function has two advantages. First, it is a strictly convex cost function, which together with the linear catch function in the model gives a strictly concave objective function. This is important because strict concavity is a necessary condition for convergence of the variables in the model to their equilibrium values (Flåm, 1993). Second, by choosing a value for $\omega = 0.01$, we end up with a marginal cost of fishing effort that can be considered constant for all practical purposes.

The single period payoffs

When the purse seine fleet chooses the level of effort and FADs to deploy in a given period t, it also automatically decides on the amount of bigeye and yellowfin bycatch to take depending on the stock sizes of bigeye and yellowfin. Also, that choice affects the unit cost of fishing per unit weight of fish caught, either by increasing the catchability coefficient, $q_{i,a}$, or by reducing the unit cost of fishing by purse seines, k_i, or both. For this paper, we assume this effect will be felt through a reduction in the cost of fishing.

The single period profit of a given player, $p = $ (PS, LLS, LLD) is then given by

$$\pi_p = \sum_i \sum_{a=0}^{A} v_p w_a q_{p,a} n_{a,t} e_{p,t} - \left(\frac{1}{1+b}\right) k_p e_{p,t}^{1+b} \tag{10.6}$$

Where $w_{a,i}$ is the weight of tuna of age a of species i and all other variables and parameters are as earlier defined.

We want here to isolate and focus attention on interactions between the players at the level of the resource. Therefore, the profit function above is formulated so as to exclude the possibility for interactions between the players in the marketplace. First, a constant price means a competitive market for fish, where the quantity put on the market by any single player does not affect the price. Second, the profit function of player, p, is assumed to depend only on the player's own effort.

The non-cooperative sequential game

There are two stages in this game. In the first stage, the purse seine fleet chooses its fishing effort, including the number and size of FADs to set, to maximize its discounted economic rent, or payoff, from the fishery over the time horizon of the game. In stage two of the game, the two longline fleets choose their fishing effort, targeting bigeye and yellowfin tuna, to maximize their payoff given the decision of the purse seine fleet in stage one of the game.

In this case, the problem of player p is to find a sequence of effort, e_t^p ($t = 1,2,\ldots,T$) to maximize its own discounted resource rent:

$$\mathrm{M_p}(n, e_p) = \sum_{\mathrm{t=1}}^{\mathrm{T}} \delta_p^t \pi_p \left(n_t, e_{p,t}\right) \tag{10.7}$$

subject to the stock dynamics given in equation (10.1) and the obvious non-negativity constraints. In the equation above, $\delta^p = (1 + r^p)^{-1}$ is the discount factor. The variable n (n_t) is the post-catch stock matrix in number of fish; and r^p denotes the discount rate of player p. The non- cooperative management scenario is what is playing out currently and therefore can be considered the status quo scenario.

The cooperative sequential game

Here, the number of purse seine and longline fishing efforts, including the number of FADs in the case of purse seines, is chosen to maximize the joint payoff from all fleets.

The goal of the cooperative players is to find a sequence of effort, e_t^p, and stock level, $n_{a,i,t}$ to maximize the joint objective functional, prof$_{com}$, given below. The cooperative management objective functional translates into maximize:

$$\mathrm{M}(n, e_p) = \sum_{p=1}^{P} \sum_{i=1}^{I} \sum_{t=1}^{T} \delta_p^t \pi_p \left(n_t, e_{p,t}\right) \tag{10.8}$$

subject to the same constraints mentioned under non-cooperation. Practically, joint management here means that in the first stage of the game, the fishing effort and the number of FADs to be employed in catching skipjack is chosen to maximize the net benefit from all fleets targeting all three species, and not just the purse seine net benefit as under non-cooperative management. In principle then, the key question here is how many FADs should be deployed, and hence how much juvenile bigeye and yellowfin catch should or should not be taken in order to maximize the joint objective functional expressed in equation (10.8) above?

To solve the model, we apply a numerical procedure whose mathematical formulation is developed in Flåm (1993), and applied in Sumaila (1995) and Armstrong and Sumaila (2001).

The data

Note that for the sake of scaling, fleet sizes of 10 purse seines fishing 200 days per year and 50 longline vessels that can set 2,000 hooks for 200 days in a year are used. The costs of operating these fleets are estimated at $44 and $40 million per year, respectively, for purse seines and longlines when the purse seines employ FADs (Reid *et al.*, 2003). Without FADs, the cost of fishing by purse seines per catch is estimated to double while the cost of fishing using longlines is assumed to decrease by 50%. The price per tonne of skipjack, yellowfin, and bigeye are estimated to be $1,500, $2,500, and $3,000 in 2008, respectively. The data and assumptions on costs and prices have been varied in order to explore how changes in these would affect the outcomes of this analysis. This is necessary because prices of up to $7,000 have been quoted recently for bigeye, for example.

The biological parameters, α and γ, are set equal to 0.8 and 1 per million tonnes, respectively, to give catches that are close to current catch levels under non-cooperative management. The survival rate of skipjack is assumed to be 0.2 for age groups 0 and 1, and 0.6 for age groups 2 to 4. Survival rates for yellowfin are 0.5 for age zero and 0.8 for age groups 1 to 5. In the case of bigeye, the survival rates are assumed to be 0.6 for all age groups (Pallares *et al.*, 2005). The initial numbers of the three different tunas are set to provide catch levels that are close to current catches under non-cooperative management. The proportion of mature skipjack and the age-dependent weights of skipjack are given in Table 10.1.

Tables 10.2 and 10.3 present the catchability coefficients applied in the model when purse seiners fish with and without FADs for the three gear types and for the three tuna species in our study.

The results

Here, we've run the status quo non-cooperative scenario and only one cooperative scenario, namely, in the extreme case where FADs are not used at all and the bycatch of juvenile yellowfin and bigeye is assumed to be zero.

Table 10.1 Parameter values used in the model

Age a	*Proportion mature*			*Weight of fish (kg)*		
(years)	*Skipjack*	*Yellowfin*	*Bigeye*	*Skipjack*	*Yellowfin*	*Bigeye*
0	0	0	0	0.6	0.6	0.6
1	0	0	0	4	4	4
2	1	0.5	0	11.3	11.3	11.3
3	1	1	0.5	22.9	22.9	22.9
4	1	1	1	38.4	38.4	38.4
5		1	1		57	57
6			1			78.1
			1			100.9

Table 10.2 Status quo catchability – current use of FADs by purse seines (noncooperation)

	Catchability (q)		
	Skipjack	*Yellowfin*	*Bigeye*
Purse seine*	0.0446	0.00218	0.002034
Deep water longline**	0	0.00272	0.003096
Shallow water longline***	0	0.001788	0

Notes
* per fleet of 10 PS boats.
** per fleet of 50 LLS boats.
*** per fleet of 50 LLD boats.

Table 10.3 No FADs catchability – (cooperation)

	Catchability (q)		
	Skipjack	*Yellowfin*	*Bigeye*
Purse seine*	0.05352	0	0
Deep water longline**	0	0.002992	0.005984
Shallow water longline***	0	0.002861	0

Notes
* per fleet of 10 PS boats.
** per fleet of 50 LLS boats.
*** per fleet of 50 LLD boats.

Table 10.4 presents the average annual net present value and catch of the different stocks of tuna taken by the different fleets under non-cooperation (purse seining with FAD) and under cooperation (without FAD).

We see from the Table 10.4 that less skipjack is caught under non-cooperation than under cooperation as purse seines now fill their capacity with only this stock. Similarly, the catches of yellowfin and bigeye are higher under cooperation than under non-cooperation, for the simple reason that under the former more juveniles of both stocks are now allowed to mature before they are fished. However, the discounted profit from fishing skipjack is lower under cooperation than under non-cooperation management, while those for yellowfin and bigeye are higher.

The average annual discounted profit from skipjack is $98 million less under cooperation. On the other hand, the average annual discounted profits from yellowfin and bigeye are higher under cooperation by about $162 and $95 million, respectively. Hence, our estimate of the net gain from all three tunas, under cooperation, is $159 million per year on average. These numbers imply that delaying cooperation by 10 years, for example, will result in a projected loss of close to $1.6 billion at present value. The average annual discounted profit

Table 10.4 Average annual net present value (NPV) and catch taken by the different fleets under cooperative and non-cooperative management

	Skipjack		*Yellowfin*		*Bigeye*	
	Average annual NPV ($m y^{-1})	*Catch ('000 tonnes)*	*Average annual NPV ($m y^{-1})*	*Catch ('000 tonnes)*	*Average annual NPV ($m y^{-1})*	*Catch ('000 tonnes)*
Non-cooperative (with FAD)						
Purse seine	885	895	168	143	95	73
Deep water longline	0	0	36	14	181	97
Shallow water longline	0	0	125	81	0	0
Cooperative (no FAD)						
Purse seine	787	1070	0	0	0	0
Deep water longline	0	0	41	16	371	161
Shallow water longline	0	0	450	246	0	0
Total non-cooperative	***885***	***895***	***329***	***238***	***276***	***160***
Total cooperative	***787***	***1070***	***491***	***262***	***371***	***161***

to the purse seine fleet, under non-cooperation, that is, when FADs are used is $1,118 million per year. Under cooperation, the discounted profit is $787 million per year. Hence, this fleet will lose about $361 million per year without the benefit of using FADs. On the other hand, the shallow and deep water fleets will gain a total of $520 million annual profits. These results imply that the purse seine fleet can be compensated for their potential loss from the non-use of FADs and still leave a surplus from cooperation of about $160 million per year.

Two other important variables, i.e. amount of effort employed by each fleet and the stock sizes of skipjack, yellowfin, and bigeye under cooperative and non-cooperative management were computed. Our calculations show that the fishing effort exerted by all three fleets increases under cooperation compared to under non-cooperation. We find that relative to non-cooperation, the purse seine effort increases by about 9% under cooperation. The equivalent increases for the shallow and deep water fleets are 86% and 32%, respectively. In a sense, this is a unique result from game theoretic models of fishing. Usually, the effort levels employed by players under non-cooperative management are higher than under a cooperative regime. The reason for this result here is the strong effect of the use of FADs on the total biomass of yellowfin and bigeye under non-cooperation. Also, the fact that by avoiding the incidental catch of juvenile bigeye and yellowfin, the purse seine fleet needs to work harder, targeting skipjack, to achieve full capacity utilization, thus increasing its fishing effort.

Our results show that the stock size of skipjack, currently not considered overfished, will decrease by 8% under cooperation, mainly because the purse fleet will have to fill up their capacity with only skipjack in the absence of juvenile bigeye and yellowfin bycatch. On the other hand, the biomasses of bigeye and yellowfin see moderate increases of between 2% and 3% under cooperation.

Sensitivity analysis on the key parameters of the model such as prices, cost of fishing, discount rates, and catchability coefficients show that changes in these parameters would affect the results generated but only in a quantitative manner. This means that the specific numbers reported in this paper should be used with caution. However, the opposing nature of the impact of changes in these parameters suggests that, on balance, our numbers are not completely off.

Concluding remarks

The goal of this study is to explore the cooperative and non-cooperative management of skipjack, yellowfin, and bigeye tuna in the WCPO, with a view to isolating the negative economic effects of juvenile bycatch by the purse seine fleet. This study shows that there is a significant gain to be made by reducing the use of FADs and the capture of sizable quantities of juveniles by the purse seine fleets active in the WCPO. This gain is estimated be about $160 million per year. To motivate owners of purse seines to agree to a significant reduction in the bycatch of bigeye and yellowfin by reducing the use of FADs, an institutional arrangement is needed to allow domestic countries using purse seines to share in the gains from cooperation.

What is interesting with the findings of this study for policy makers is that we can indeed have a win-win-win outcome by putting in place a cooperative management regime for the WCPO tunas. In the first place, our analysis suggests that there would be an overall increase in the discounted economic rent for the fisheries. Secondly, we show that there would be no need to cut fishing effort under cooperative management, implying that the social loss that normally accompanies cuts in fishing effort is not expected to apply in this case. Indeed, the model predicts the need for more hands in the fishery. Finally, the stock sizes of both yellowfin and bigeye show mild increases under cooperation while that of skipjack suffers a loss, but this stock is generally considered not to be overfished.

11 Impact of management scenarios and fisheries gear selectivity on the potential economic gains from a fish stock[1]

Introduction

The objective of this chapter is to undertake bioeconomic analysis of Namibia's hake fishery to support optimal sustainable management. Among the species of hakes inhabiting the Namibian exclusive economic zone (EEZ), cape hake and deep-water hake are of major importance to the fishery. These two species are so identical in appearance that they are often treated as one and the same (Wysokinski, 1986). Both species are relatively long-lived, reaching ages of up to and over 9 years. Hakes are usually found close to the bottom of the water during daytime but rise to intermediate water during night-time, probably following their prey. Here, we study hake as if it were a single stock.

The management of Namibian hake consists of two main processes. First, a process of determining the annual total allowable catch (TAC), and second, a process that allocates the TAC among a number of license holders who employ different fishing gears to exploit hake. These two steps are carried out by the Ministry of Fisheries and Marine Resources, Namibia (MFMR), using inputs from scientists, industry, and management. It is anticipated that the results of this study will provide insights that would help enhance the work of the MFMR with respect to both the determination and allocation of the TAC for hake.

The study focuses sharply on three important characteristics of the hake fisheries. One, the fact that wetfish and freezer trawlers, the two main vessel types used to exploit the resource, have different fishing grounds and consequently target different age groups of the hake stock. Two, the fact that the two vessels land hake in forms that influence the price they receive per unit weight of their catch. Three, each vessel group has its own cost structure, and hence land hake at different costs per unit weight.

The work in this chapter fits into the general literature on the economics of shared stocks (see for instance, Munro, 1979; Levhari and Mirman, 1980; Fischer and Mirman, 1992; Sumaila, 1997b; Armstrong 1998). Sumaila (1997b) is a study of the North-east Atlantic cod in the Barents Sea. This is a fishery located in the Northern hemisphere, which has been very well studied. On the other hand, the present chapter studies the Namibian hake fishery, which is based in the less developed South. This fishery has not been well studied, especially with respect to bioeconomic analysis, and therefore serves as a greater challenge to the modeler.

For instance, while there are many studies that look into the selectivity patterns of the coastal and trawler vessels active in the Barents Sea (see for example, Armstrong *et al.*, 1991; Larsen and Isaksen, 1993), there are hardly any that have looked closely at the selectivity patterns of the wetfish and freezer trawlers active in Namibia's EEZ. In comparison to Sumaila (2000), this paper is more ambitious because it incorporates stock recruitment and dynamics, and seeks to advise not only on how much of a predetermined TAC should be allocated to the two vessel groups (as was the objective in Sumaila, 2000) but also on the overall size of the TAC.

In the next section, I briefly discuss the hake fishery, I then present the bioeconomic model, including the data used for the computations. Following the numerical results of the study are presented. One key finding is that a management strategy for hake that seeks to protect either the juvenile or mature part of the stock from exploitation makes good economic sense. This result may indeed be one explanation for the recent surprise decline in Namibian hake stocks, which followed the introduction of a policy of 60:40 share of the hake TAC to the wetfish and freezer fleets, respectively.

The Namibian hake fishery

The hake stocks are one of the three most important fish species of the highly productive Namibian EEZ. The others are horse mackerel (*Trachurus trachurus*) and pilchard (*Sardinops ocellatus*). The main reason for the high productivity of the Namibian EEZ is the Benguela upwelling system prevalent in the coastal zone of Namibia and other Southem African countries.

Hake catches reached a maximum of over 800,000 tonnes in 1972, averaging some 600,000 tonnes annually during the period from the late 1960s to mid-1970s. As expected, this period of high catches was followed by lean years, with average catches of less than 200,000 tonnes from the mid-1970s to 1980. This, however, rose again and remained relatively stable between 300,000 and 400,000 tonnes for most of the 1980s. It is stated in Hamukuaya (1994) that during those years of high catches there was a large proportion of young fish between the ages of 2 and 3 years old, probably accounting for the low catches in later years. Bonfil *et al.* (1998) show that due to the high catches of hake, horse mackerel, and pilchard attributable to the activities of distant water fleets prior to independence, Namibia inherited a fishery well below its productive potential.

It is worth mentioning that the fishing sector is important to the economy of Namibia, with the hake fisheries being an important part of this. According to the MFMR, hake contributed about N$230 million or 7.4% of Namibia's estimated exports in 1994.

The model

The fishing fleets targeting hake

A variety of fishing vessels are used to catch hake; differing in their gross registered tonnage, engine horse power, processing equipment, and freezing capacity. However, the bulk of hake are landed by wetfish and freezer trawlers. For instance, in 1994 out of a total of 108,213 tonnes of hake landed, 99,152 tonnes were by wetfish and freezer trawlers. This is well over 90% of the total landings of hake that year. The rest is landed using monksole trawlers, longliners, and mid-water trawlers (Moorsom, 1994; Sumaila, 2000). Data from 1995 and 1996 show that the dominance of the bottom trawlers in the hake fisheries continues unabated (Ministry of Fisheries and Marine Resources, 1996). As a result of the overwhelming dominance of the bottom trawlers in the demersal hake fishery, I focus my attention on these vessels and organize the wetfish and freezer trawlers into two separate and distinct entities assumed to be managed by two different bodies, from now on, to be known as the Wetfish Industry Group (W) and the Freezer Industry Group (T), respectively.[2] These two groups are assumed to interact under (i) command, (ii) cooperative, and (iii) non-cooperative environments, as explained later in the chapter.

Recruitment and stock dynamics of Namibian hake

The Beverton–Holt age-structured model forms the basis for modeling the biology of hake in this study. According to Punt (1988), this model corresponds closely to the stock biomass observed in International Commission for the South-East Atlantic Fisheries (ICSEAF) Divisions 1.3 and 1.4 (which are parts of the Namibian EEZ) from 1956 to 1985, the parameters of the model having been estimated using results of virtual population analysis.

Let the spawning biomass, B_t^s, be defined by the following equation:

$$B_t^s = \sum_{a=0}^{a_{\max}} p_a w_a n_{a,t} \tag{11.1}$$

where $a = 0,1, \ldots, a_{\max}$, denotes age group a hake; $a_{\max}$ is the last age group; w_a stands for weight of hake of age a at the start of the year; $t = 1,2, \ldots, T$, is fishing years, with Tdenoting the last period; p_a stands for the proportion of age a hake that is mature; and $n_{a,t}$ represents the number of age a hake in year t.

The stock–recruit relationship, R_t, is given by:

$$R_t = n_{0,t} = \frac{aB_t^s}{(\alpha\beta + B_t^s)^{-\gamma}} \tag{11.2}$$

where $n_{0,t}$ is the number of recruits in year t; and α, β, γ are parameters of the extended Beverton–Holt stock–recruit relationship (Punt, 1988).

From the above, the basic stock biomass can be represented by the equations below:

$$n_{a,t} = \theta n_{a-1,t-1} - h_{a,t}, \quad \text{for } 0 < a < A$$

$$n_{A,t} = \theta n_{A,t} - \theta n_{A-t,t-1} - h_{A,t}, \quad n_{a,0} \text{ given} \tag{11.3}$$

The function $h_{a,t} = \sum_{p} q_{p,a} n_{a,t} e_t$ denotes the total catch by both players of age group a hake in fishing period t; θ is the age independent natural survival rate; e_t is the fishing effort exerted on cod in period t, while q stands for the catchability coefficient of the hake harvesting vessels. The reader should note that the stock dynamics of the last age group of hake is given special treatment. This is meant to capture the fact that all age $a_{\max}$ hake do not die at the end of a given period.

On selectivity and catchability

To determine the appropriate catchability coefficients to apply in the model, the method outlined in the Appendix is employed. A key input to the method is gear selectivity.[3] For a well-studied fishery such as the Barents Sea cod fishery, it is easy to find these from the literature, but this is not the case for the Namibian hake fishery. Therefore, to form an opinion on the selectivity patterns of W and F, a number of fisheries people in Namibia were interviewed. A clear consensus that came out of the interviews was that the wetfish trawlers (because their fishing grounds are close to the shore) target mainly young fish while the freezer trawlers target mainly mature fish, because they operate further into the sea. Using this background information, it is assumed in the model that wetfish trawlers exploit age groups 1 to 6 hake, while freezer trawlers target age groups 5 to 9.[4] The selectivity pattern for hake reported in Punt and Butterworth (1991) is used to set a total overall selectivity for each age group. Hence, the sum of the selectivity by the two vessel groups on a given age group is equal to the selectivity for that age group, reported in Punt and Butterworth (1991).

Economics of the hake fisheries

As mentioned earlier, the MFMR is assumed to manage the hake stock for the benefit of Namibia as a whole. It therefore acts as a sole owner who seeks to obtain maximum economic benefits from the resource without destroying the resource base. We determine an equilibrium outcome which I term the "command outcome" to depict the behavior and actions of the MFMR. In this outcome, the MFMR decides both the TAC and its allocation to the two parties, in a manner which will ensure maximum total economic benefit from hake. Two other equilibrium outcomes to be computed are the non-cooperative and cooperative. The former is determined to serve as a benchmark for comparison with the cooperative and command outcomes. In addition, it serves as the "threat point" when the Nash cooperative solution is determined (see Nash, 1953; Munro, 1979).

For two reasons, it is assumed in this paper that the price per unit weight of hake faced by both players is perfectly elastic. The first reason relates to the fact that the Namibian supply of hake is not big enough to influence the international market for hake under normal circumstances. Secondly, the focus here is on the impacts of gear selectivity stemming from interactions at the level of the stock, not at the level of the market.

The catch cost function of a given player p in period t, $C(p,t)$, is modeled as an "almost" linear function of its fishing effort, $e_{p,t}$ (see Sumaila, 1995):

$$C\left(e_{p,t}\right)\frac{k_p e_{p,t}^{1+b}}{1+b} \tag{11.4}$$

where $b = 0.01$, and $k_p/(1+b) \approx k_p$ is the cost of engaging one fishing fleet for one year.

Let the single period profit of player p be given by:

$$\pi_{p,t} = \pi_p\left(n_t, e_{p,t}\right) = v_a \sum_{a=0}^{A} w_a q_{p,a} n_{a,t} e_{p,t} - C\left(e_{p,t}\right) \tag{11.5}$$

where $n_{a,t}$ is the age- and period-dependent stock size in number of fish; w_a is the mean weight of fish of age a; and $q_{p,a}$ is the age and player dependent catchability coefficient, that is, the share of age group a hake being caught by one unit of fishing effort by player p.

The non-cooperative scenario

Under this scenario, it is assumed that there is no regulator coordinating the actions of the two fleets. Furthermore, there is no possibility for credible communication between W and F: the management of each fleet takes the actions of the other as given, and chooses its own strategies to maximize own discounted economic rent. That is, each player finds a sequence of effort levels, $e_{p,t}$, so as to maximize its discounted economic rent:

$$M_p\left(n, e_p\right) = \sum_{t=1}^{T} \zeta_p^t \pi_p\left(n_t, e_{p,t}\right) \tag{11.6}$$

subject to the stock dynamics given by equations (11.2) and (11.3) above and the obvious non-negativity constraints. In the equation above, $\zeta_p = \left(1 + r_p\right)^{-1}$ is the discount factor. The variable $n(n_t)$ is the post-catch stock matrix (vector) in number of fish; and r_p denotes the interest rate of player p.

The command scenario

Here, the commander (or regulator), which in this particular case is the MFMR, seeks to find a sequence of effort, $e_{p,t}$, and stock levels, $n_{a,t}$, to maximize a weighted average of the objective functionals of the two fleets, denoted prof_{com}. β and $(1-\beta)$ indicate how much weight is given to the own objective functional of W and F by the commander. For a given $\beta \in [0, 1]$, the cooperative management objective functional translates into maximize

$$\text{prof}_{\text{com}} = \beta M_1(n, e_1) + (1-\beta) M_2(n, e_2) \qquad (11.7)$$

subject to the same constraints expressed by equations (11.2) and (11.3). The important point to note here is that the MFMR chooses the β which produces the highest total economic rent. This then determines both the overall TAC and how much of this should be caught by W and F, respectively. After determining these, the MFMR simply issues a directive, which we assume the fishers are under the obligation to comply with.

The cooperative scenario

Under this scenario, too, there is no commander, W and F work together freely and cooperatively to determine a TAC and its allocation to themselves. The key point to note at this juncture is that the outcome agreed upon must be an incentive compatible with their own interests (see Binmore 1992). In other words, the outcome and hence the payoffs to each player must be at least as much as what the player will receive if he or she decides not to cooperate.

The two players may choose to work for a cooperative "with" or "without" side payments arrangement. The latter refers to a situation in which all players want to participate in actual fishing, and thus will not accept any compensation not to do so. The former is the opposite of this: all possible solutions are considered, including the possibility of buying out a player. Given the definition of the command scenario in this paper, the solution to the cooperative "with" side payments is close to the "command" outcome. In both cases, the objective is to maximize the weighted average of the objective functions of the two fleets under the appropriate constraints. The main difference between the two is in the way the gain from cooperation is shared. In the case of the command scenario, the commander decides this, while under cooperative with side payments, a rule based on an application of the Nash bargaining scheme (Nash, 1953; Munro, 1990) is used.[5]

The solutions to the model are pursued numerically (see Flåm, 1993), rather than analytically, for two reasons. First, the complex age-structured nature of the model makes it analytically difficult to solve (see Conrad and Clark, 1987). Second, the objective of the current paper is to produce quantitative rather than qualitative results.

Table 11.1 Values of parameters used in the model. Maximum age, weight, taken from Punt and Butterworth (1991). Catchability coefficients derived, initial stock size, and proportion mature estimated

Age a (years)	*Selectivity Sa*	*Catchability coefficient*		*Proportion mature (pa)*	*Weight w(a) (kg)*	*Initial numbers (millions)*
		W	*F*			
0	0	0	0	0	0.001	2
1	0.007	0.00672	0.0060	0	0.0345	1.3
2	0.032	0.00307	0.0162	0	0.0935	0.64
3	0.216	0.0207	0.0162	0	0.187	0.4
4	0.426	0.0384	0.0162	0.5	0.319	0.28
5	0.972	0.05759	0.0004	1	0.55	0.18
6	1.028	0.0580	0.0060	1	0.929	0.13
7	1	0	0.0162	1	1.445	0.1
8	1	0	0.0162	1	2.108	0.04
9	1	0	0.0162	1	2.542	0.03

Model data

The biological, economic, and technological data are mostly taken from Punt and Butterworth (1991), Punt (1988), Sumaila (2000) and the MFMR. Table 11.1 displays (i) the proportion mature of each age group, p_a; (ii) the average weight, w_a; (iii) the total selectivity for each age group, S_a; (iv) the initial numbers of each age group of fish; and (v) the catchability coefficients for each vessel type. The latter are calculated by splitting the total selectivity according to the observed targeting patterns of juvenile and mature hake by the two vessels; and using the framework in Appendix 1 of Sumaila (1997b), The rest of the model parameters are given the values: $\alpha = 6300$ (million) $\beta = 0.16$; $\gamma = 1.0$ (Punt, 1988); $a_{\max} = 9$ (Punt and Butterworth, 1991). The natural survival rate, θ, is assumed to be 0.81 per year. Price per kilogram for the landings of the wetfish ($v_1 =$ N\$[6] 8.18) and freezer ($v_2 =$ N\$7.38) trawlers are taken from Sumaila (2000). The costs of employing the wetfish and freezer trawlers for one year are determined from data from the Namibian fishing industry to be N\$12.29 and N\$39.90 million, respectively. A discount factor of 0.952 (equivalent to a real interest rate of 5%) is assumed.

The results

Payoffs in a fully economic setting

By a fully economic setting I refer to a situation in which the fisheries manager incorporates all the appropriate economic parameters and variables (prices, costs, and discount factors) into the decision-making process of how to manage the resource.

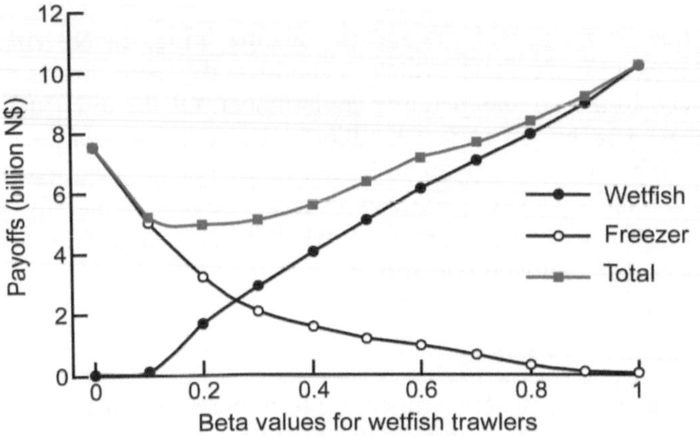

Figure 11.1 Payoffs to wetfish, freezer fleets separately and jointly in the fully economic setting.

Figure 11.1 displays the discounted economic rent achievable under cooperation for different β-values. This graph shows how the payoffs obtained by using wetfish and freezer trawlers change with varying β-values, that is, with changing emphasis on the preferences of the wetfish fleet relative to those of the freezers.

The best discounted economic rent computed under the command, noncooperation and cooperation regimes are reported in Table 11.2. This table shows that under the fully economic environment, the command and the cooperative with side payments outcomes give a total discounted economic rent of N$10.23 billion over the 25-year time horizon of the model. To achieve this, all the TAC should be taken by the wetfish trawler fleet (that is, when $\beta = 1$; see Figure 11.1). Under this scenario, we see that protection of the mature stock by reducing the freezer fleet catch to zero turns out to be bioeconomically sensible. Following the sharing rule mentioned earlier, the wetfish and freezer fleets receive N$7.18

Table 11.2 Total discounted economic rent (N$billion) under the different management regimes and assumptions of the economic environment

Management regime	*Command*			*Cooperative*			*Noncooperative*		
	Wetfish	*Freezer*	*Total*	*Wetfish*	*Freezer*	*Total*	*Wetfish*	*Freezer*	*Total*
Fully economic	10.23	0	10.23	6.18	0.96	7.14	4.63	0.50	5.13
Cost-less labor input	13.27	0	13.23	8.15	1.32	9.47	6.75	0.90	7.65
Equal price, fully economic	0	7.52	7.52	4.47	0.88	5.35	3.54	0.54	4.08

and N$3.05 billion dollars, respectively, in the cooperative with side payments scenario.

The Nash cooperative "without" side payments outcome brings in N$7.14 billion (when $\beta = 0.6$, see Figure 11.1), which is significantly more than the N$5.13 billion produced in the non-cooperative environment. Of the total, the wetfish fleet pulls in N$6.18 billion (N$4.63 billion under non-cooperation), and the freezer fleet brings in N$0.96 (N$0.50 billion under non-cooperation). In comparison to the command and cooperative scenarios, the non-cooperative outcome is very bad – it produces an economic rent which is only about 50% of what is achievable under the command scenario.

Payoffs in a cost-less labor input setting

The motivation for implementing this scenario comes from observations I made during my fieldwork: key decision-makers in the MFMR were of the view that given the high unemployment level in Namibia, the government is more concerned with providing as many sustainable jobs in the fishing sector of the economy as possible. I interpret this point in this model to imply that the alternative cost of fishing labor inputs is taken to be zero by the fisheries managers.

In Figure 11.2, the discounted economic rents determined under the cooperative scenario, for different β-values, are presented. In addition, Table 11.2 reports the best results under cooperative, command, and non-cooperative scenarios, respectively.

From this Table 11.2 we see that the command outcome produces a payoff of N$13.27 billion. This happens when the wetfish fleet alone catches the stock, that is, when the preferences of the wetfish fleet are given full weight by management ($\beta = 1$). A payoff of N$9.47 (wetfish: N$8.15 and freezer: N$1.32) billion is realized under cooperation "without" side payments. Here, cooperation with side

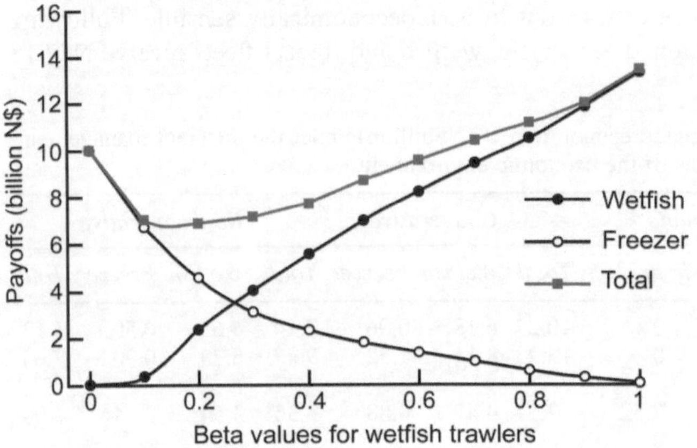

Figure 11.2 Payoffs to wetfish, freezer fleets separately and jointly in the cost-less fishing labor input setting.

payments results in payoffs of N$9.56 and N$3.71 billion for wetfish and freezer trawlers, respectively. Finally, non-cooperation leads to a total payoff of N$7.65 (wetfish: N$6.75 and freezer: N$0.90) billion.

The good outcomes achieved by the wetfish fleet relate to the fact that they enjoy a number of "private" advantages. First, their landings receive, on average, a higher price per unit weight than those of freezer trawlers (see Sumaila, 2000). Second, the proportion of labor cost to total fishing cost is higher for the wetfish than the trawler fleet. Thus, in the cost-less labor input scenario, the performance of the wetfish fleet improves further. Third, this class of fishing vessels appears to have an advantage in that it targets juvenile fish and can, therefore, undermine the freezer fleet in a competitive situation.

To find the impact of the higher price received by the wetfish fleet, the model is re-run under the assumption that landings by the wetfish fleet receive the same price per unit weight as landings by the freezer fleet. Figure 11.3 displays the discounted economic rent achievable under cooperation in a fully economic setting. This graph shows that in this case it is optimal to let only the freezer fleet do the catching. From Table 11.2, we see that when both fleets face the same price, the command outcome gives N$7.52 billion.

Standing biomass

Table 11.3 presents the average standing biomass and the catch size and proportion, over the 25-year time horizon of the model. A comparison of the numbers under the two management scenarios reveals the following. One, the command or cooperative with side payments scenario produces the best possible health for the stock under both assumptions of the economic environment. Two, the non-cooperative situation is terrible for the health of the stock, producing average standing biomasses which are well below those attained in the command and cooperative with side payments scenarios. Three, the

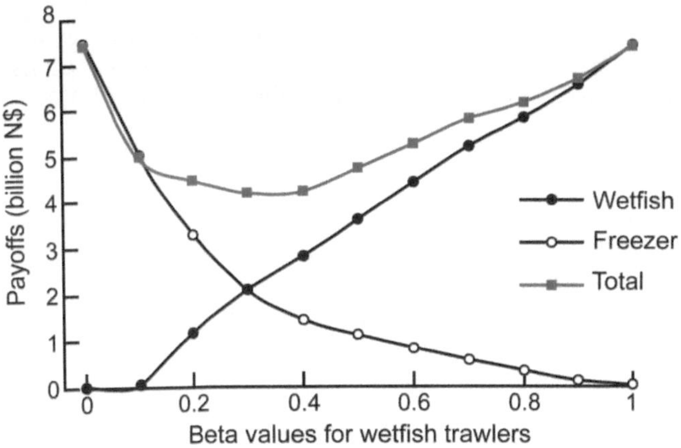

Figure 11.3 Payoffs to wetfish, freezer fleets separately and jointly, when both vessel types face the same price.

Table 11.3 Average standing biomass, catch (thousand tonnes) and proportion of catch by the wetfish trawlers

(a) Base case scenario

Management regime	*Fully economic*		*Cost-less labor*	
	Biomass	*Catch (proportion)*	*Biomass*	*Catch (proportion)*
Command	1330	129 (100%)	1330	141 (100%)
Cooperative	1300	85 (95%)	1300	96 (95%)
Non-cooperative	917	87 (95%)	896	99 (95%)

(b) Equal price scenario

Management regime	*Fully economic*		*Cost-less labor*	
	Biomass	*Catch (proportion)*	*Biomass*	*Catch (proportion)*
Command	1690	79 (0%)	1330	122 (100%)
Cooperative	1280	73 (95%)	1300	85 (95%)
Non-cooperative	938	80 (94%)	913	92 (94%)

cooperative without side payments scenario is second best, as it mitigates against the biological waste shown to exist in the non-cooperative scenario, but falls short of the *optimum optimorum* achievable under cooperation with side payments or the command scenario.

A comparison of the outcomes under the different assumptions of the economic environment indicates that: under the command and cooperative scenarios, the same average standing biomass is achieved under the two economic environments. On the other hand, under non-cooperation because lower cost of fishing labor inputs implies a greater "race" for the fish: lower cost pushes the equilibrium stock size lower. Hence, a policy that tends to assume away the cost of fishing will also tend to lower the average standing stock size. The reader should note that qualitatively the "no price difference" scenario produces results that are similar to those discussed in the above paragraphs (see Table 11.3b).

Catch sizes and proportions

The average catch and the proportion of the catch in the base case (no price difference) scenario are reported in Table 11.3a (Table 11.3b). It is worth noting that the catch sizes for the various scenarios are good indicators of both the number of boats and labor required to land the catch. In fact, one may assume a linear relationship between catch and these input variables. Hence, we do not discuss separately the labor required to take the landings predicted under the different scenarios.

A number of observations can be made from Table 11.3a (Table 11.3b). First, in the fully economic environment, an average catch of 87,000 (80,000) tonnes is obtained under non-cooperation. The average catch under the command and cooperative without side payments scenarios are 129,000 (79,000) and 85,000 (73,000) tonnes, respectively. Second, the cost-less fishing labor input assumption results in higher catch under all the scenarios. However, the gains in catch under the non-cooperative scenario come at a biological cost – the average standing biomass is lower than in the fully economic scenario.

The optimal catch proportion for the wetfish trawlers ranges between 95% and 100%, except when the same price is assumed for the landings of the two vessel types. In which case, a catch proportion of zero for the wetfish fleet is found to be optimal under the command and cooperative without side payments scenarios. These numbers are clearly different from the current policy of 60%:40% in favor of the wetfish fleet.

Discussion and concluding remarks

The study shows that the choice and implementation of management strategies for hake can have huge effects on the bioeconomic benefits from the resource. To illustrate this point, take the estimated average annual catch predicted by the study: a wide range of between 73,000 and 141,000 tonnes, depending on the management scenario and the assumptions underlying the economic environment. This calls for careful analysis on the part of the MFMR to guide its decision-making process. Clearly, with proper data, models such as the one presented here can produce useful insights for practical management of the hake fisheries of Namibia.

An important conclusion that can be derived from the results of this study is that a management policy that seeks to protect either the juvenile or mature part of the stock from exploitation produces good bioeconomic outcomes. This is because in all cases the best outcomes are achieved either when only the wetfish or freezer trawlers are allowed to exploit the resource. This result is particularly interesting because it may well be one reason for the surprising decline in the hake stock size after about 3 years of the introduction of a policy of 60:40 division of the hake TAC between the wetfish and freezer trawlers.

Another point to be made from the findings of the paper is that cooperation, whether it comes about through negotiations or is enforced by a controller, can lead to significant economic gains to both parties. Furthermore, the study shows that the need for good data, both biological and socio-economic, cannot be over-emphasized. In addition, studies to find out the selectivity patterns of the vessels used to exploit not only hake but other important species in Namibian waters, would be very useful.

Finally, it is worth mentioning that the study is, as with all modeling and computational exercises, partial in some sense. For instance, the current model does not explicitly capture inter- and intra-species interaction.

12 Managing bluefin tuna in the Mediterranean Sea[1]

Introduction

Following a general global pattern (e.g. Pauly *et al.*, 2002; Worm *et al.*, 2009), the Atlantic bluefin tuna (BFT) stock is currently at risk of being overfished to depletion. The widely-accepted primary reason for the current state of this stock is its common property and shared stock status, which together can easily drive exploiters of a given natural resource into non-cooperative behavior (Munro, 1979), known as the "tragedy of the commons" (Hardin, 1968).

To deal with the common-property and shared stock problem of tunas and tuna-like species including BFT in the Atlantic Ocean and adjacent areas, the International Commission for the Conservation of Atlantic Tunas (ICCAT) was established in 1969. One of ICCAT's major responsibilities is to set and allocate BFT's catch quotas according to its scientific stock assessment. However, ICCAT has consistently set the quotas much higher than the levels recommended by its scientists since 1995 (ICCAT Reports, 1994, 1995, 1996, 2005, 2006, 2007, 2008a, 2008b; MSBN, 2004; BBC News, 2007; Renton, 2008). Thus, the organiszation has been harshly criticized for its failure to manage BFT sustainably (MSBN, 2004; BBC News, 2007; Renton, 2008). Consequently, some more drastic and immediate actions are called for, including a complete shut-down of the fishery; listing BFT on the Convention on International Trade in Endangered Species of Wild Fauna and Flora (CITES); and cutting the current annual catch quota by more than half. In order to evaluate these actions and improve BFT stock sustainability, this chapter provides a background review to the fisheries and management regime in the Mediterranean Sea, analyses why management has failed, and then proposes policy changes to address this failure.

The fisheries

The Atlantic BFT, native to both the West and East Atlantic Ocean, can be naturally divided into two groups: West[2] and East Atlantic BFT, which differ both in their habitat and their life histories. Both groups of BFT are highly migratory and have a long life span of up to 30 years. In terms of fisheries, the East Atlantic BFT stock supports the Eastern Atlantic Ocean area and the Mediterranean Sea. In this chapter, BFT fisheries in the Mediterranean Sea are the main focus.

The BFT fishery in the Mediterranean Sea started in the 7th Millennium BC (Desse and Desse-Berset, 1994). The popularity of Japanese sushi and sashimi worldwide during the 1980s made the BFT much more economically attractive than before (Fromentin and Ravier, 2005; Porch, 2005). For example, a single BFT was auctioned in the Tokyo market for US$396,700 in 2011.[3] Consequently, vessel capacity, vessel power, and new storage innovations for BFT experienced tremendous increases in the 1980s and 1990s, which imposed severe pressure on the BFT stock.

Bluefin tuna fisheries and stock status

Figure 12.1 illustrates the BFT historic catch by gear type in the Mediterranean Sea from 1950 to 2005. This figure shows that from the 1950s to the early 1970s, total catches were stable at around 5,000 to 8,000 tonnes per year. Starting from the early 1970s, large changes were observed with catch peaking in the mid-1970s, followed by an unusual drop by the early 1980s. From then on to the mid-1990s, the catches increased steadily from 9,000 to 40,000 tonnes per year. After that, there was a substantial decrease in catch to 24,000 tonnes per year in the most recent decade, which seems to serve as an indication of effective management. However, instead of official catch reductions, this drop is regarded by the ICCAT Standing Committee on Research and Statistics (SCRS) to be due to underreporting (ICCAT, 2008b).

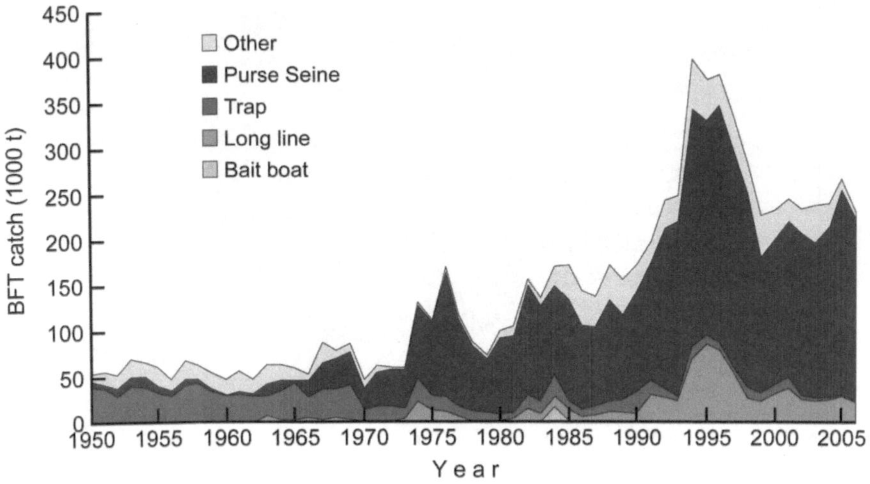

Figure 12.1 BFT catch in the Mediterranean Sea.
Source: ICCAT Report 2008a.

Figure 12.1 also shows some interesting patterns in the catch by gear type. The bait boat fishery, which mostly catches juvenile fish, contributes very little to the total catch. The long line catch peaked in the mid-1990s along with the purse seine catch. The trap catches have consistently declined over time and now have totally disappeared. In contrast, catches from the purse seiners has been consistently increasing over time, which makes purse seines the major gear used to catch BFT in the Mediterranean Sea currently.

According to ICCAT SCRS, this unusually high increase in purse seine catches is related to the growth of BFT fattening farms, since the purse seine is the best gear type for ensuring the capture and transfer of live tuna. It is estimated that only 200 tonnes of Mediterranean BFT were "consumed" in farms in 1997, while between 20,000 to 25,000 tonnes had been fattened in farms every year since 2003 (ICCAT, 2008b). In fact, as a consequence of the huge expansion of purse seine fleets, no spawning refuge seems to exist for BFT in the Mediterranean Sea anymore because almost every inch of the sea is now covered by fishing effort (ICCAT, 2008b).

Figure 12.2 displays the pattern of catch at age in the Mediterranean Sea from 1955 to 2006. The catch of age 0 BFT has decreased since the 1960s and is barely observed today. The catches of other age groups all increased in weight in 2006 compared to 1950. Relatively, the total weight share of age groups in 1950 is different from that in 2006, which more or less reveals that the current stock structure in fish numbers has changed a lot compared to what it used to be a few decades ago.

Increasing BFT catches have led to rapid stock declines over the years. According to the stock assessment analyses reported by ICCAT, the decline

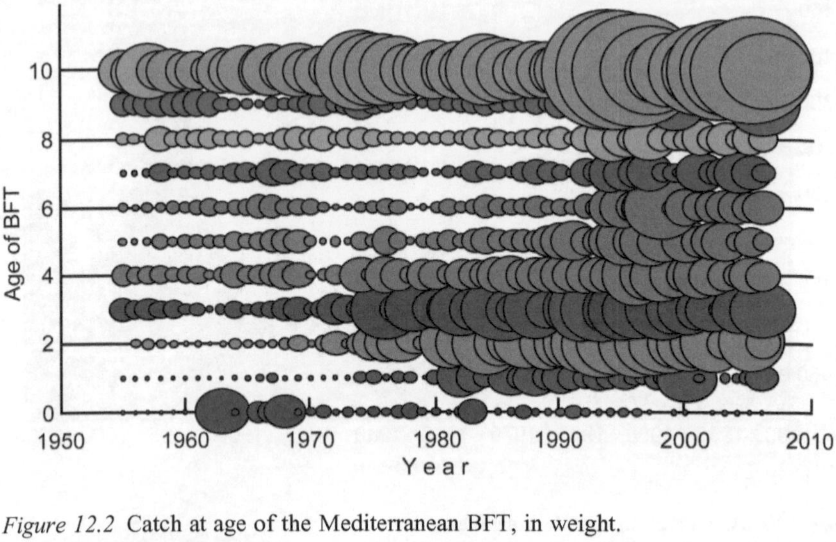

Figure 12.2 Catch at age of the Mediterranean BFT, in weight.
Source: ICCAT Report 2008a.

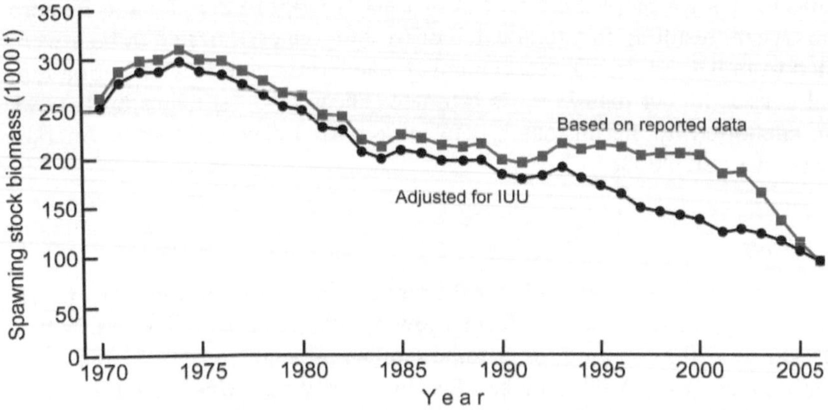

Figure 12.3 Spawning stock biomass.
Source: ICCAT Report 2008a.

of spawning stock biomass (SSB), one of the most important indicators of stock abundance and health, is evident from analyses on catch data. Figure 12.3 shows the estimated SSB from 1970 to 2005. In this figure, two model predictions, based on reported and adjusted catch data, respectively, are presented. The adjusted catch data takes illegal, unreported, and unregulated catch into account. Both of these two model runs show that, except for a slight increase in the period from 1970 to 1974, SSB has declined persistently, with current SSB estimated to be only 40% of its peak in 1974.

Illegal, unreported, and unregulated fishing

Illegal, unreported, and unregulated (IUU) fishing is widely recognized as one of the biggest concerns for BFT management in the Mediterranean Sea and other Atlantic Ocean areas. WWF (2006) found huge gaps between national reports on BFT trade and official catch reports to ICCAT, indicating that a large amount of IUU fishing takes place in the region. The cited study estimated that the total BFT catches in the Eastern Atlantic Ocean and the Mediterranean Sea, recorded through international trade, were approximately 45,000 tonnes in both 2004 and 2005, which was 40% above the total allowable catch (TAC) of 32,000 tonnes set by ICCAT. If the catches by national fleets in Spain, France, and Italy for domestic markets were also included, the total catches could be well above 50,000 tonnes per year. The same study determined that EU (mostly French) and Libyan fleets are largely responsible for most of the IUU catches (WWF, 2006).

ICCAT is also fully aware of this IUU problem. In 2006, based on the number of vessels operating in the Mediterranean Sea and their catch rates, ICCAT estimated total catches to be close to 43,000 tonnes in the Mediterranean Sea in the early 2000s. In 2008, a new evaluation by ICCAT suggested a 2007 total catch of 47,800 tonnes for the Mediterranean Sea and 13,200 tonnes for the Eastern Atlantic Ocean, resulting in a total catch of 61,000 tonnes. These numbers were estimated from ICCAT's BFT vessel number, catch rates, and stock information; the total is even higher than WWF's estimate. These IUU estimates by ICCAT are also supported by the mismatch between reported data and various market sales data (ICCAT, 2008b).

BFT farming

After BFT is caught wild and alive with purse seine, farms are used to fatten them in floating cages for periods from a few months to up to 1–2 years. WWF (2004) estimated that about 21,000 tonnes of wild-caught tuna were put into BFT farm cages in the Mediterranean Sea in 2003, which was around 66% of the declared TAC. In fact, detailed farming data are pretty scarce; only a few countries' figures are available. According to WWF (2004), 975 tonnes, 1,180 tonnes, 3,980 tonnes and 1,400 tonnes of wild-caught BFT were put into farms in Croatia, Spain, Italy, and Turkey, respectively, in 2002.

It is important to note that current BFT farming is different from traditional farming, i.e. aquaculture, which consists of a complete production chain from hatcheries to feeding and to harvests. In contrast, BFT farming only fattens wild BFT. Since BFT is highly migratory and requires different environmental conditions during its different life stages, it will be difficult to have a complete farming chain for BFT (Susannah, 2008). Some scientists estimate that at least 10 years are needed to get BFT to breed via land-based hatcheries. However, many scientists are even skeptical of this, due to the complex nature of BFT behavior and life history (Susannah, 2008). Some claim that BFT fattening would help solve the overfishing problem, but I beg to disagree: fattening is bound to impact the stock abundance of BFT negatively because much more fishing effort will be targeting juvenile BFT as result. The other concern arising from BFT farming is that highly dense farms, which are common, might also have undesirable environmental impacts: one from leftover bait, which has negative impacts on tourism, and the other from tuna processing without disposing wastes (Miyake *et al.*, 2003). Further, the use of chemicals and medicines (e.g. hormones, antibiotics) in the baits is a concern for food safety and quality, which is faced by all other aquaculture industries.

Economic benefits of bluefin tuna

BFT is considered a "culture-specific" product because most of the world's consumption occurs in Japan with over 45 countries competing to supply this market (Carroll *et al.*, 2001). The Mediterranean region is one of the major exporters of BFT to Japan. In this section, the key economic indicators related to BFT stocks in the Mediterranean Sea are estimated, including the

Table 12.1 Gear specific BFT ex-vessel prices

Gear	*2006 Prices (USD/kg)*
Longline/trap	10.67
Purse seine	9.44
Others	16.14

Source: NMFS, 2010.

total landed values, the total fishing costs, the resource rent, employment number supported by the fishery, and added values through the BFT fish value chain.

Total landed value

In order to calculate the total landed value, information regarding BFT catches and ex-vessel prices is needed. Table 12.1 shows gear specific price data for the Atlantic BFT obtained from NMFS (Susannah, 2008). The ex-vessel BFT price for longline or trap is around US$10.67 per kilogram, while catch by purse seine is sold at US$9.44 per kilogram.

Using gear specific prices and gear specific catch data from ICCAT (2008b), the total BFT landed values are computed for countries targeting tuna in the Mediterranean Sea, presented in Table 12.2.

Table 12.2 shows that around US$49.9 million of landed BFT value were captured by the countries in the MENA region and US$176.9 million by non-MENA region countries in 2006. Tunisia records the highest landed value among MENA region countries while France captures the highest landed value among all countries.

Total costs of BFT fishing

Corresponding to landed values are fishing costs. BFT fishing costs have two components: variable and fixed costs. Furthermore, variable costs include fuel, repair, other operation costs, and labor costs. Fixed costs are composed of depreciation costs, payment to capital, and other fixed costs. Here, purse seine fishing costs and revenue data from Concerted Action (2006, 2007) are used to compute the percentage of total fishing costs relative to revenue.[4] Then this percentage is assumed to hold for BFT in the Mediterranean Sea, and this is then used to estimate the BFT fishing costs. According to Concerted Action (2006, 2007), the fishing costs relative to revenue percentages are 99.6% for Spain, 87.6% for France, 73.7% for Italy, 96.6% for Portugal, 99.8% for Korea Republic and 85.0% for Taiwan. For those countries whose data are missing, the average figure is used for the Mediterranean area, which is 90.4%. The costs estimated for each country, also presented in Table 12.2, show that Tunisia has

Table 12.2 Mediterranean BFT landed value and resource rent estimates in 2006

Country/entity	*Total reported catch (t)**	*Landed value (thousand US$)*	*Total cost (thousand US$)*	*Resource rent (thousand US$)*	*Unit resource rent US$/kg*
MENA region					
Algeria	1,038	10,555	9,539	1,016	0.98
Israel	0	0	0	0	0
Libya	1,280	12,255	11,075	1,180	0.92
Morocco	190	3,047	2,754	293	1.54
Tunisia	2,545	24,045	21,729	2,316	0.91
Regional total	5,053	49,902	45,096	4,806	
Non-MENA region					
China	0	0	0	0	0
Croatia	1,022	9,648	8,719	929	0.91
EC Cyprus	110	1,174	1,061	113	1.03
EC Spain	2,689	26,259	26,143	116	0.04
EC France	7,664	73,862	64,681	9,181	1.20
EC Greece	254	2,497	2,257	240	0.94
EC Italy	4,694	46,673	34,417	12,256	2.61
EC Malta	263	2,806	2,536	270	1.03
EC Portugal	11	117	113	4	0.36
Japan	556	5,933	5,362	571	1.03
Korea Rep.	26	277	276	1	0.04
Panama	0	0	0	0	0
Serbia & Montenegro	0	0	0	0	0
Taiwan	5	53	45	8	1.60
Turkey	806	7,609	6,876	733	0.91
Yugoslavia Fed.	0	0	0	0	0
Regional total	18,100	176,908	159,872	24,422	
Total	23,153	226,810	197,582	29,228	

Data source: ICCAT, 2008b.

the highest fishing costs in the MENA region and France has the highest in the non-MENA region.

Resource rent

Resource rent is defined here as the landed value (gross revenue) minus fishing costs. The estimated resource rent for each country is also included in Table 12.2. The total resource rent is estimated to be about US$4.8 million for the MENA region (9.6% of the landed value) and US$24.4 million for the non-MENA region (13.8% of the landed value) in 2006. Thus, the non-MENA resource rent is about 5 times the rent accruing to the MENA countries. Tunisia and Italy are the two countries with the highest resource rent, among MENA region, and all the countries, respectively. In the same table, the unit resource rent is also reported and Morocco and Italy are found to have the highest figures in the MENA and the non-MENA regions, respectively.

Institutional setting

International Commission for the Conservation of Atlantic Tunas (ICCAT)

ICCAT was created to manage more than 30 tuna and tuna-like species in the Atlantic Ocean and adjacent seas, including the Mediterranean BFT. The Commission, composed of 48 Contracting Parties (countries/political entities),[5] is a Regional Fisheries Management Organisation (RFMO) responsible for combining a wide array of scientific and socio-economic information into setting TACs of Atlantic tuna species. The quota set by ICCAT is then split among member countries who are individually responsible, but not obliged, to manage their fleet in accordance with the annual TAC. ICCAT is also responsible for collecting and analysing statistical information and making recommendations.

Determination of TACs by ICCAT

ICCAT is responsible for setting the TACs based on scientific evidence. Stock assessment analyses are performed by ICCAT SCRS, who are responsible for providing scientific advice to ICCAT on the TAC and quota allocation among member countries. However, ICCAT has traditionally set much higher TACs than recommended by this Committee.

The comparison between scientifically recommended TACs and actual TACs set by ICCAT is given in Table 12.3, which shows a disregard for scientific advice and therefore the future health and sustainability of BFT stocks. For the year 2010, scientists estimate that even with a quota of 8,000 tonnes per year, BFT stocks have about a 50% chance of rebuilding by the year 2023, yet the

Table 12.3 East Atlantic and Mediterranean BFT annual quotas and landings

Year	*Science-based TAC recommended (t)*	*Quota set by ICCAT (t)*	*SCRS estimate (t)*
2003	15,000	32,000	>50,000
2004	15,000	32,000	>50,000
2005	15,000	32,000	>50,000
2006	15,000	32,000	>50,000
2007	15,000	29,500	61,000
2008	15,000	28,500	34,120
2009	8,500–15,000	22,000	–
2010	8,000	19,950	–

Data source: ICCAT, 2006; 2007; 2008a; 2009; 2010.

TAC set by ICCAT was nearly 70% above scientific recommendations (ICCAT, 2009, 2010).

Allocation of quota among countries

After setting TAC, ICCAT allocates shares of the annual TAC to its Contracting Parties. How the shares are divided has undergone changes in two different periods. From 1983 to 1991, ICCAT allocated the TACs among countries mainly according to their historical catches. In addition, the spatial distribution of stock, and the proximity to coastal states, especially in small and developing countries, have also been taken into consideration (Grafton *et al.*, 2006). However, CPs (Contracting Parties) without large historical catches argued for changes in the allocation formula in the 1990s and succeeded in persuading ICCAT to increase their share in 2001 (Grafton *et al.*, 2006). The allocated quota is transferrable among member countries, though transfers have to be made under the approval of ICCAT.

Table 12.4 presents the allocation of the BFT quotas to different countries/groups targeting the East Atlantic BFT stock. The quotas remained almost constant from year 2003 to 2006. Among non-EC countries, Morocco received the highest portion of quota, followed by Japan.

Furthermore, Table 12.5 shows the allocation among EU countries, but only for 2004 and 2005. Three countries – Spain, France, and Italy – received about 55% of the TAC in the East Atlantic and the Mediterranean Sea in 2004 and 2005.

Compliance enforcement

In order to help carry out the objectives of ICCAT, CPCs (Contracting Parties & Cooperating non-Contracting Parties, Entity, and Fishing Entity) collect scientific

Table 12.4 BFT quotas (*t*) allocated by ICCAT

Country/entity	*2003*	*2004*	*2005*	*2006*
Algeria	1,500	1,550	1,600	1,700
China	74	74	74	74
Croatia	900	935	945	970
European Community	18,582	18,450	18,331	18,301
Iceland	30	40	50	60
Japan	2,949	2,930	2,890	2,830
Tunisia	2,503	2,543	2,583	2,625
Libya	1,286	1,300	1,400	1,440
Morocco	3,030	3,078	3,127	3,177
Others	1,146	1,100	1,000	823
Total	32,000	32,000	32,000	32,000

Source: Council Regulation Nos 2287/2003, 27/2005.

Table 12.5 BFT quota (*t*) allocation among EU countries

Country/entity	*2004*	*2005*
Greece	326	323
Spain	6,317	6,277
France	6,233	6,193
Italy	4,920	4,888
Other	654	650
Total	18,450	18,331

Source: Carroll *et al.*, 2001; NMFS, 2010; Concerted Action, 2006, 2007; Teh and Sumaila, 2011; Dyck and Sumaila, 2010; Pontecorvo *et al.*, 1980; ICCAT Reports, 2009 and 2010; Grafton *et al.*, 2006; Council Regulation Nos 2287/2003 and 27/2005.

data and report to SCRS by 31 July of each year. However, since no penalty is associated with this data reporting, partial, late, or even no data are often submitted.

CPCs are obliged to establish a high seas international enforcement system. Until 1997, there was no at-sea boarding or inspection. However, a Port Inspection Scheme was established in 1997 to inspect both flag and non-flag state vessels during off-loading and transhipment in ports. Consequently, a list of vessels believed to be engaging in IUU fishing was published in 1999. In contrast, according to ICCAT's 1998 and 2000 recommendations, a list of fishing vessels was authorized in 2002.

CPCs are also responsible for enforcing compliance through domestic policies. Records of non-compliance are considered by the ICCAT Compliance Committee, trade restrictions, or revoking of vessel registration may follow. For non-Contracting Parties, the Permanent Working Group for the Improvement of ICCAT Statistics and Conservation Measures (PWG) is responsible for overseeing and collecting their information.

Domestic BFT management

Although much of the focus of tuna management in the Mediterranean Sea is on the actions of ICCAT, its yearly TAC is only a recommendation, with implementation left to the individual member states. Currently, ICCAT members are not known for managing their shares of the tuna TAC using tradable permits or individual transferable quotas (ITQs). It appears that the majority of ICCAT members fishing in this area use licensing systems to manage their fisheries.

While there are attempts at effort control by several nations,[6] lack of effective management at the national level is likely a reason behind the dramatic decline of BFT stock in the Mediterranean. In 2007, three countries – Italy, Spain, and France – landed more than 17,800 tonnes over their quota of BFT

(Bregazzi, 2007). Additionally, it is estimated that Italy, Spain, and Libya, were responsible for under-reporting their catches of BFT by more than 16,000 tonnes in 2007 (Bregazzi, 2007).

Why has the current institutional framework failed?

Shared fish stock

There is a general consensus that common shared fish stocks, which include transboundary fish resources found in more than one exclusive economic zone (EEZ) of countries, highly migratory species in multiple EEZs or high seas, or fishes in the high seas (Munro *et al.*, 2004), are difficult to manage (Munro, 1998; Munro *et al.*, 2004; Payne *et al.*, 2004). Since targeting commonly shared fish stocks usually leads to inevitable externalities, i.e. fishing by one country influences the stock and thus fishing in the other countries, management of shared fish stocks requires countries to cooperate, which is very difficult to achieve. To solve this problem, game theory is often applied to examine the cooperative incentives among different entities to find win-win solutions. However, since the benefits of cooperation are always highly uncertain, it is extremely challenging to reach agreements in practice.

BFT is a typical shared fish stock since it is highly migratory, crossing multiple EEZs and the high seas. Therefore, it shares all the challenges of managing shared fish stocks, which by nature needs a very high level of cooperation and enforcement. Not surprisingly, the current ICCAT regime, with low monitoring and loose enforcement, cannot succeed in preventing the overfishing of BFT stocks without significant improvement.

Conflicts between members and non-members

Non-ICCAT members can also fish BFT, which forms another big barrier to the successful management of ICCAT. According to Miyake (1992), significant amounts of catches are taken by non-ICCAT countries. Officials from the Japan Fisheries Agency pointed out that catches by non-member countries may be more than 80% of those by member countries (Miyake, 1992). An increasing number of boats have been reported flying flags of non-member countries to avoid regulation. This large proportion of catches taken by non-ICCAT countries serves as a significant barrier for effective management of the ICCAT quota system. This barrier, together with the highly shared nature of BFT, results in a significant level of IUU catches of BFT in the Mediterranean Sea.

Subsidies

BFT overfishing is exacerbated by government subsidies, which are financial transfers, direct or indirect, from the public sector to the private sector (Sumaila *et al.*, 2010). Subsidies in the Mediterranean BFT fisheries can be divided into two connected groups: (1) subsidies for fleet modernization; and (2) subsidies to BFT farms. In the following, the current situation of these two types of subsidies is described.

A tremendous expansion of BFT and other tuna farming activities in the Mediterranean Sea have been observed recently. However, it is believed that EU subsidies are the main underlying reason for the expansion (WWF, 2004). It is reported that in some countries, e.g. Spain, the market price for farmed tuna in 2003 was well below the production cost of tuna fattening farms (WWF, 2004).

EU companies get subsidies mainly through the Financial Instrument for Fisheries Guidance (FIFG), which aims for "fleet renewal and modernization of fishing vessels" and "aquaculture development," "processing and marketing of fishery products" and others.[7] FIFG helps to build and modernize purse seine fleets and plays an important role in the Mediterranean tuna fattening expansion. Besides FIFG subsidies, matching funds from national and regional administrations are usually available depending on domestic policies. It has been roughly estimated that at least €19–20 million of EU public funding has contributed to the tuna farm expansion (WWF, 2004). These subsidies covered up to 75% of the fleet and farm investment cost (Council Regulation No. 1451/2001). In Spain alone, this subsidy amounted to €6 million. Although the total subsidy value for fleet modernization is unclear, available evidence shows that huge amounts of public funding have been involved. For example, 40 powerful high-tech French purse seine vessels were known to have been modernized with subsidies (WWF, 2004). These subsidies directly encourage overfishing in the Mediterranean Sea, which is another important reason why the current institutional framework is ineffective. Unfortunately, ICCAT has failed to address this issue.

Policy recommendations

Here, alternative policy schemes and recommendations are provided to ensure the sustainable exploitation of BFT in the Mediterranean Sea.

Institutional improvement in ICCAT

TAC reduction

It is clear from the data analyses that ICCAT needs to substantially reduce the current TAC by following scientific advice. A US National Marine Fisheries Service study showed that if ICCAT had not raised the TAC from 1,160 to 2,660 tonnes in 1983, the adult population would have been 3.4 times what it was in the early 1990s (Powers, 1992).

In order to reduce the TAC, a higher level of cooperation needs to be established among BFT fishing countries/entities. It is expected that the reduction of TAC can be beneficial for all the participants if they cooperate in the management and conservation of BFT. More research should be carried out to determine the economic benefits of multilateral cooperation among participants and to discover acceptable compensation mechanisms. For example, if the TAC is heavily reduced, small-scale coastal fisheries may lose profits in some countries while large-scale fisheries may benefit in other countries. In this case, ICCAT can set up platforms for contracting members to negotiate with each other to reach agreements such that with the countries that benefit most compensate those who suffer losses. ICCAT also needs to make its members aware of how large the potential benefits from cooperation are and thereby motivate them to cooperate. A mutual compensation fund can be established to enable such cooperation among countries. This fund can help cover some of the costs of an effective inspection program, proposed below.

At-sea inspection and alternating scrutiny system

A functioning and effective Reporting and Monitoring (R&M) system is very pivotal to the success of compliance enforcement. Thus, ICCAT needs to establish a much more strict R&M system. Currently there is only port boarding and inspection. Instead ICCAT could establish an at-sea boarding or inspection program at the international level. In addition, local ICCAT member countries could develop an alternating peer scrutiny system, i.e. if there are three countries: A, B, C; then A could inspect B, B inspects C and C inspects A. This design can avoid co-deviation: if A gets to scrutinize B and B scrutinizes A, they might have the incentive to collaborate and underreport each other's catches.

Penalty regime

The reason why ICCAT cannot succeed in combating IUU fishing is that it lacks an effective detection and penalty system (Sumaila *et al.*, 2006). Since there is no penalty for overfishing, the economic incentives for reducing harmful practice are almost zero. Thus, ICCAT could establish and enforce a penalty system. When an IUU event is found, penalties have to be paid by the country responsible for this IUU fishing. The funds raised from this penalty program can be used for stock rebuilding, research and for covering R&M costs.

Seeking legal rights to manage non-ICCAT entities

Currently, ICCAT has no mandate to manage non-ICCAT entities, which not only adds a significant amount of catches to the total catch, but also imposes negative externalities on ICCAT members. Furthermore, entities do not have economic incentive to become ICCAT members since non-ICCAT entities are free from any restrictions. Thus, ICCAT can seek legal rights to manage non-ICCAT entities. For example, political pressures in the UN or trade restrictions might be potential routes to achieving this.

Subsidies reduction in the EU

As described earlier in this contribution, EU subsidies have become a threat to maintaining sustainable BFT stocks since they have largely distorted investment decisions for fleet modernization and farm expansion. If BFT is managed well, the EU will be the largest beneficiary since they have the largest quota in the Mediterranean Sea. Thus, possibilities exist that ICCAT can induce EU to remove harmful subsidies and use the saved resources on programs to reduce overcapacity and overfishing.

Marine protected areas

To cope with the management of the shared fish stock, marine protected areas (MPAs) might be a useful policy instrument (Salm and Clark, 1989; Halpern and Warner, 2002). MPAs are areas in the ocean within which human activities are regulated more stringently than elsewhere (Sumaila and Charles, 2002). Currently the world has more than 5,000 MPAs.[8] As recognized by many, MPAs conserve biodiversity, protect tourism and cultural diversity, increase fish productivity and provide insurance against stock collapse (Kelleher, 1999). Due to these benefits, MPAs are generally proposed as a tool for effective fisheries management if the targeted species are not highly migratory or have relatively fixed spawning sites.

It is well documented that BFT migrate to well defined areas to spawn (Cury, 1994; Fromentin and Powers, 2005; Fromentin, 2006; OCEANA, 2008), which is supported by Block *et al.* (2001), who studied BFT migration behavior using tag data. Because BFT congregate to spawn, they are highly vulnerable to commercial fishing at their spawning times (Alemany *et al.*, 2010), which makes MPAs a potentially effective management instrument. ICCAT needs to fully consider the potential of MPAs as one of the regional management tools to ensure sustainable management of BFT in the Mediterranean Sea. In order to investigate whether MPAs are effective management tools for BFT, more research should be carried out by ICCAT to learn how BFT migrates over spaces, and to determine BFT spawning grounds, etc. With such information and additional economic analyses, locations and sizes of MPAs can be intelligently decided (Halpern, 2003).

Listing in Convention on International Trade in Endangered Species of Wild Fauna and Flora as an endangered species

As ICCAT consistently shows its inability to effectively manage BFT, conservationists have appealed to other alternative authorities, especially CITES, which is an international body with an objective to “ensure that international trade in specimens of wild animals and plants does not threaten their survival.” So far, the listing of BFT in CITES has been proposed twice, in 1992 by Sweden and in 2010 by Monaco.[9] However, Sweden withdrew the proposal in 1992 and the proposal in 2010 got denied, both due to feverish rejection by some ICCAT member countries, in particular, Japan. Thus, listing in CITES Appendix I is a difficult path and seems infeasible in the near future. As stated before, other more feasible management tools could be used to manage BFT under the current circumstances.

Domestic management

Individual transferable quota (ITQ) system

The individual fishing quota system, which involves allocating TAC share to individuals or firms with restrictive monitoring, is one of the economically effective management tools at our disposal currently (Costello *et al.*, 2008). As of 2008, about 10% of global marine catch was managed by ITQs (Chu, 2009). Since ICCAT has allocated TAC to each country, it is possible for them to adopt domestic ITQ systems. However, besides the usual problems in regular fisheries: equity (who gets the quota) and highgrading (smaller fish are discarded) issues, BFT ITQ implementation has more challenges. First, BFT is highly migratory, so it is easy for IUU fishing to occur. Second, BFT is a fish resource that is shared by multiple countries, which highly decreases the incentives of these countries to comply with the TACs.

Dedicated Access Privileges (DAP) program

With the Dedicated Access Privileges (or Limited Access Privileges) program, individuals, communities, or others are granted the privilege of catching a portion of the TAC or commercial quota. DAP is different from ITQs in two ways. First, individuals and communities or other groups are also eligible to receive fishing rights. Second, it grants the privilege to fish, not property rights. As mentioned above, ITQs are often criticized for privatizing public resources; DAP, instead, avoids this problem by only renting out fishing rights. Therefore, BFT fishing countries can consider adopting DAP as their domestic management strategies.

Optimal resource allocation: a case study of Tunisia

Given the total quota allocated and other countries' actions, individual countries may have possibilities to improve their domestic management. They can choose whether to sell the quota directly, or sell BFT after fishing or fattening them. Optimally allocating quota shares to these different activities can improve a country's total net benefits from its allocated quota. In this section, Tunisia is used as a case study to illustrate an optimal quota allocation of BFT among the choices of selling the quota to another country, consuming fish domestically, and using the catches as inputs to farms. The analysis is carried out with a simple economic model:

$$\pi_t = v_q * qs_1 * Q_t + \sum_s v_s * H_s - \text{Cost\_}H_t + v_A * qs_3 * Q_t * (1+G) - \text{Cost\_}M_t \tag{12.1}$$

$$\text{Cost\_}H_t = (c_1 + c_2)(qs_2 + qs_3)Q_t \tag{12.2}$$

$$\text{Cost\_}M_t = c_3(qs_3 * Q_t * (1+G))^2 \tag{12.3}$$

$$qs_1 + qs_2 + qs_3 \leq 1 \tag{12.4}$$

Equation (12.1) describes the profit from fishing and farming BFT. Generally speaking, t is the time, s is the gear, Q represents quota and v is the price. c_1 to c_3 are the cost coefficients and qs_1 to qs_3 are the quota shares. The first component, $v_q * qs_1 * Q_t$, is the revenue from the selling quota, v_q is the quota price, Q_t is the total quota for year t and q_1 is the percentage of the sold quota relative to the total quota, Q_t. The second component, $\sum_s v_s * H_s - \text{Cost\_}H_t$, is the net profit for fishing. Since the BFT price is gear specific, each price by gear, v_s, is multiplied by its corresponding catch H_s, and then their fishing cost is deducted, modeled by equation (12.2), in which c_1 is the fixed cost and c_s is the variable cost. The catch consists of $qs_2\ Q_t$ directly for consumption and $qs_3 Q_t$ for farming. The last component $p_A * qs_3 * Q_t * (1+G) - \text{Cost\_}M_t$, calculates the farming profit. Here, $qs_3 * Q_t * (1+G)$ is the weight after fattening and v_A is the average price for farmed fish. The cost of BFT farming is modeled by equation (12.3). To comply with the TAC, the sum of qs_1, qs_2, and qs_t is not greater than 1, which is the constraint represented by equation (12.4).

In this model, apart from the quota selling price, all the price and cost parameters in the profit function of quota selling, directly consuming, and farming are known.[10] Then, if the quota selling price varies between US$8 and US$10 per kilogram, what the corresponding quota allocations for these three options would be is examined. Figure 12.4 shows the result for the sensitivity analysis.

Figure 12.4 shows that as the quota selling price increases, there are different combinations of quota proportions that can optimize the total profits for Tunisia. When the price is lower than US$8.2 per kilogram, the quota should not be sold to other countries, but rather used domestically. Here, the quota proportion for direct consumption is similar to that for farming. As the quota prices go up, it is more profitable to sell the quota to another country instead of fishing themselves, and the farming proportion should be larger than the share for direct consumption.

This figure illustrates that a country can improve the allocations of quota to optimize its total profit given its fixed quota. Similarly, sensitivity analyses can also be conducted by varying other price and cost parameters. Since each country has its own different parameter values, which will change the model results quantitatively, each country needs to take different strategies based on its own situation.

It is worth noting that this model is only a simple illustration of individual countries' spaces of resource optimization. It is conditional on many assumptions: for example, the prices, costs and profit structure are deterministic, IUU fishing is limited. In reality, the problems are much more complicated. If there are lots of unreported catches, the quota system will be defeated and not effective. Consequently, there will be no basis for this kind of resource optimization.

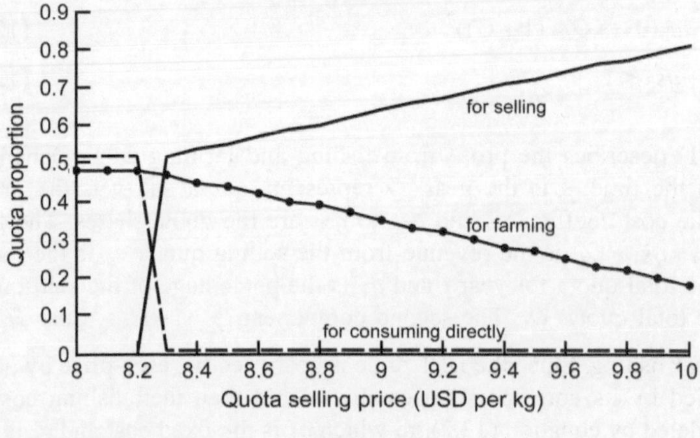

Figure 12.4 Optimal quota allocation.

In this sense, the improvement of this type of domestic management will highly depend on the implementation of other international regulations on BFT.

Conclusions

In this chapter, the fisheries and stock status of BFT in the Mediterranean Sea and related management issues are reviewed:

(1) The spawning stock biomass of BFT has decreased by 60% from its 1974 quantity;
(2) The total BFT catch per year in the Mediterranean Sea is about 24,000 tonnes in recent years. However, IUU in the same area could be as high as 47,800 tonnes. Purse seine is currently the major gear used to catch BFT, which is largely associated with BFT farm expansion in the region;
(3) The total landed value for Mediterranean BFT is estimated to be US$226.8 million a year, which results in US$29 million of resource rent. It is also estimated that about 3,500 full-time fishing jobs are supported by BFT stocks and this fishery has a multiplier effect on national economies of about US$635 million;
(4) ICCAT has consistently set TACs above the level recommended by scientists.

As pointed out in the analysis, many factors prevent successful management of BFT. Among them, the common-property and shared stock nature of the fishery, the existence of non-ICCAT members and EU fishery subsidies are all important factors. In order to address these issues, ICCAT is suggested to strengthen institutions by developing effective cooperative mechanisms, introducing enforceable penalty regimes and reporting/monitoring systems.

In addition, ICCAT needs to seek ways to manage non-ICCAT members and convince the EU to reduce their fishery subsidies for BFT fattening farms, and vessel modernization. The implementation of MPAs is also recommended to support regional management, and it is suggested that individual countries use a Dedicated Access Privileges program and resource optimization to improve their domestic management.

Appendix
Theoretical basis of the solution procedure

Classification, assumptions, and limitations

The objective of this appendix is to give a discussion of classification, assumptions, and limitations of game theory. A treatment of these issues is considered important because, in the first place, the choice of refinement of the Nash equilibrium idea typically depends on the type of game under study and the environment in which it is played. Secondly, a simple classification of games plus a discussion of its assumptions and limitations would put the content of this book in the right perspective. The plan is to treat two different forms of classifications (this does not imply that there are only two forms of classifications of games; the reason only two are discussed is that they are sufficient for our purposes) and their underlying assumptions, and then to follow this up with a brief discussion of the limitations of game theory. The two classifications of interest are classifications of games into (i) *cooperative* and *non-cooperative* games, and (ii) games of *complete* and *incomplete* information.

The standard way to classify games is, usually, into non-cooperative and cooperative games. Cooperative games are further divided into cooperative games *with side payments* and those *without side payments*. A side payment is a sum of money which may be paid by one player to another in order to facilitate or hinder the signing of a pre-play contract (Binmore and Dasgupta, 1986b).

The assumptions underlying the choice of strategies in a formal game played cooperatively, are, (i) the players may communicate costlessly and without restriction, (ii) the players may enter any agreements whatsoever that they choose, and (iii) mechanisms exist for enforcing such agreements; the latter assumption is, by and large, the most important and tricky.

Non-cooperative games cover a wide variety of possibilities but attention is usually focused on what Harsanyi calls "*tacit games*" and which Binmore and Dasgupta (1986b) prefer to call "*contests*." Here, "contests" and "tacit games" are used interchangeably. Before the commencement of tacit games, the players are assumed not to have any opportunities, explicit or implicit, for any type of communication at all.

It was von Neumann and Morgenstern (1944), who first classified games into games of complete information and those of incomplete information. Games of complete information are considered basic, since those of incomplete information can be reduced to those of complete information using the theory of Harsanyi (1982). In a game of complete information, the players are assumed to know all relevant information not explicitly forbidden by the rules of the game. To be more precise, players in a game of complete information are assumed to know (Binmore and Dasgupta, 1986a):

- the rules of the game; and
- the beliefs, strategic possibilities, and preferences of the players in the game.

It is necessary to expatiate a bit more on what complete information implies, and what it does not imply. Firstly, complete information implies that information is common knowledge (Aumann, 1976). This means that, not only does each player know it, but also that each player knows that each player knows it, that each player knows that each player knows that each player knows it ... and so on *ad infinitum* (Myerson, 1984; Mertens and Zamir, 1985). Secondly, the player's beliefs about the world need not be consistent, and lastly, information need not be perfect[1] in this formulation.

Three further assumptions of game theory, listed in Binmore and Dasgupta (1986a), which characterize a game of complete information are:

- A rational player quantifies all uncertainties with which he or she is faced using subjective probability distributions and then maximizes utility relative to these distributions. It is further assumed that each of his or her probability distributions is common knowledge.
- A rational player can duplicate the reasoning process of another rational player provided he or she is supplied with the same information.
- It is common knowledge that all players are rational.

A natural question to ask at this juncture is: what do is meant by a rational player? A player is said to be rational if the player is a "well-integrated" personality, with his/her motivations precisely defined via a preference ordering, so that he or she maximizes utility given his/her subjective belief.

Proposition 1

If a rational analysis of a contest is able to single out an optimal strategy choice for each player, then this profile of strategy choices must constitute a Nash equilibrium of the game.

Proof

If this were not so, there would be at least one player who would have an incentive to choose some other strategy, implying that the strategy singled out for him or her is not his or her optimal strategy: an obvious contradiction.

Definition 1

Suppose that the players in a game are named $i = 1, 2, \ldots, N$ and A_i is the set of feasible strategies of player i. The elements of A_i may include mixed strategies. A Nash equilibrium is then a strategy profile $(a_1^*, a_2^*, \ldots, a_n^*)$ with a_i^* element of A_i such that each a_i* is a best reply for player i to the choices a_j^* by the players $j \neq i$. If $U(a_1, a_2, \ldots, a_n)$ is the expected utility that a player i derives from the use of strategy profile $(a_1, a_2, \ldots, a_n)$, then this means that

$$U_1(a_1^*, a_2^*, \ldots, a_n^*) > U_1(a_1, a_2^*, \ldots, a_n^*)$$

$$U_2(a_1^*, a_2^*, \ldots, a_n^*) > U_2(a_1^*, a_2, \ldots, a_n^*)$$

$$- - - - - - - - - - - - - -$$

$$U_n(a_1^*, a_2^*, \ldots, a_n^*) > U_n(a_1^*, a_2*, \ldots, a_n)$$

for all $a_1 \in A_1$, $a_2 \in A_2$, … $a_n \in A_N$.

From the above, it is clear that a game theorist needs to make a number of assumptions which may sometimes be absurd in real life. Depending on the type of game, the following assumptions, among others, need to be made:

- Beliefs need to be common knowledge.
- Individuals must be optimizers.
- Each person must be capable of unlimited computational ability.
- Players may communicate costlessly and without restriction.
- Players may enter any agreement and there are mechanisms for enforcing such agreements; and
- Players are rational.

The limitations of game theory derive mainly from the difficulties of meeting these assumptions in real life. For example, an experimental study by Guth *et al.* (1982), confirms that not even trained economists can be relied upon to behave "rationally" in the simplest of structured bargaining games. See Binmore and Dasgupta (1986a:10–14) for a rationalization of some of these problems.

Finally, a few words on the difference between a game-theoretic analysis on the one hand, and a behavioral analysis on the other. A game theoretical analysis would normally focus on those factors which have, or appear to have, a genuine strategic relevance to the situation being analysed. It is the contention of the game theorist that this does not include the bulk of maneuvers common in real-life negotiations such as flattery, abuse, and other subtle attempts to put the opponent at a psychological disadvantage. These factors would certainly be of importance in a behavioral analysis but have no place in a game-theoretic analysis. A rational player considers such factors as irrelevancies, since the main concern is the outcome of the game to be obtained rather than the manner in which the outcome is achieved. Much as game theorists agree that behavioral analysis is important, it still seems sensible to tackle the simple problems first, before attempting to solve the more difficult ones using behavioral analysis.

Optimization

What follows constitutes the underlying theoretical and philosophical basis for the algorithm used to compute Nash equilibria in this book. The algorithm is based on optimization using *gradient* and *Lagrangian methods*.

The single agent unconstrained problem

First, consider a one person non-constrained problem of the following form:

$$\max_{x} U(x) \tag{A1}$$

This problem can often be solved by applying the rule:

$$\dot{x} = x'(t) = \frac{dx(t)}{dt} = U'(x) \tag{A2}$$

or more generally, when U is not differentiable in the classical sense, by

$$\dot{x} \in \partial U(x) \tag{A3}$$

where $x'(t)$ is the rate of change of x with respect to time and $u'(x(t))$ is marginal utility derived by consuming x amount of stock at time t. In principle, the equation above is an expression of what is termed the *gradient method.* An interpretation of this method is that, to solve his or her problem, the person has to change his or her consumption of x with respect to time according to the magnitude and direction of his or her marginal utility. In particular, x remains unchanged (steady) when the marginal utility is zero at any time t.

The multi-agent (game) unconstrained problem

Now, suppose there are many players, who are in a conflict situation, with utility functions given by $u_i(x)$, where $x = \{x_i, x_{-i}\}$. Let the objective of each player be given by

$$\max_{x_i} U_i(x), \tag{A4}$$

Just as in the first case, the gradient method prescribes that each player i solves his or her problem by changing his or her own decision variable x_i according to the rule

$$\dot{x}_i = \frac{\partial\, U_i(x)}{\partial\, x_i}, \quad i = 1, 2 \tag{A5}$$

Where $\partial u_i(x)/\partial x_i$ is the partial derivative of $u_i(x)$ with respect to x_i.

The single agent constrained problem

So far, problems without any constraints are constructed: definitely a simplification. To ease difficulties, suppose there is only one player subject to the constraint ξ such that $x \in \xi$, where ξ is a closed convex set. It can be assumed, without loss of generality, that ξ has the representation $x \in \xi \Leftrightarrow C(x) \geq 0$, where $C: R^n \rightarrow R$ is a concave function. The single player's problem then becomes

$$\max_{\{x\}} U(x),\ x \in \xi \tag{A6}$$

Usually this class of problem is solved by applying the classical Lagrangian technique or method. The use of the classical Lagrangian method here will result in a profile of multipliers (y) that can decrease or increase according to the value of $c(x)$. For some good reasons (mainly convergence), our solution procedure is amenable only to the case where y converges monotonically to its steady state value, y^*. To converge monotonically means $c(x)$ cannot assume a value greater than 0. In other words, monotonocity here helps to simplify our convergence analysis. With the classical Lagrangian, the assumption that $y \geq 0$, means that for y multiplied by $c(x)$ to serve as a penalty for constraint violation, $c(x)$ must take negative values, otherwise, y multiplied by $c(x)$ will serve as a bonus for constraint violation – surely, this is contrary to intentions.

The important question now is: how to ensure that y converges monotonically? To answer this question, a modified version of the classical Lagrangian is introduced. First, let us define $c(x)^-$ as follows

$$C(x)^- = \min(C(x), 0), x \in \xi, C(x) \geq 0, C(x)^- = 0 \tag{A7}$$

Instead of the ordinary Lagrangian, let's define the following alternative Lagrangian for the single player

$$L(x; y) = U(x) + yC(x)^- \tag{A8}$$

Then the solution profiles for the stated problem, as prescribed by the gradient method, are given by

$$\dot{x} = \frac{\partial L(x; y)}{\partial x} \tag{A9}$$

$$\dot{y} = -\frac{\partial L(x; y)}{\partial y} = -C(x)^- \tag{A10}$$

For the case where the functions are not differentiable in the classical sense, substitute ($\in$) for ($=$) in (A9) and (A10) to obtain the solution profiles.

The above stated system ensures that y converges monotonically all the way to y^*. Let's now turn to the multi-agent constrained problem.

The multi-agent (game) constrained problem

The optimization problem in a multi-agent constrained game context takes the following form:

$$\max_{\{x_i\}} U(x_i, x_{-i}), \;\; C(x) \geq 0. \tag{A11}$$

The modified Lagrangian for this problem is given by

$$L^{-}(x; y) = U_i(x_i, x_{-i}) + yC(x)^{-} \tag{A12}$$

and the optimality conditions when the functions are differentiable are:

$$\dot{x}_i = \frac{\partial L^{-}(x; y)}{\partial x_i} \tag{A13}$$

and

$$\dot{y} = -\frac{\partial L^{-}(x; y)}{\partial y} = -C(x)^{-}. \tag{A14}$$

Or in the more general case, where the functions are not differential in the classical sense,

$$\dot{x}_i \in \frac{\partial L^{-}(x; y)}{\partial x_i}, \tag{A15}$$

and

$$\dot{y} \in -\frac{\partial L^{-}(x; y)}{\partial y} = -C(x)^{-} \tag{A16}$$

The discussion in the subsection above is relevant, where models with *i* number of players in constrained fisheries games are dealt with.

Optimization of different types of problems, both constrained and non-constrained are discussed in this Appendix. This discussion constitutes the general theoretical story behind the solution procedure applied in most of the chapters in this book. The main goal is to compute Nash non-cooperative equilibrium solutions in fisheries games. The theoretical basis for the existence, convergence, uniqueness, and optimality conditions for such equilibria has attracted the attention of many mathematical economists (e.g. Sundaram, 1989; Fundenberg and *Tirole*, 1991).

Notes

2 Game-theoretic models of fishing

1 Based on *Marine Policy*, 23(1), Sumaila, U.R., "A review of game theoretic models of fishing," 1–10, 1999, with permission from Elsevier.
2 See a discussion of this class of stocks under the section "Transboundary/migratory/straddling stock models."
3 A Nash non-cooperative equilibrium is an array of strategies, one for each player in the game, such that no player regrets his/her chosen strategy.
4 Sumaila (1996) presents the main mileposts in the historical development of the theory of games.
5 One of the earliest applications of game theory was in political science: in their paper of 1954, Shapley and Shubik used the Shapley value to determine the power of members of the United Nations Security Council. The Shapley value is a solution concept, characterized by a set of axioms that associate with each coalition game, V, a unique outcome, v. Four other early applications of game theory worth mentioning are to philosophy (Braithwaite, 1955); to evolutionary biology (Lewontin, 1961); to economics (Shubik, 1962); and to insurance (Borch, 1962).
6 This model has its shortcomings though, for example, it does not include any density-dependent self-regulating mechanisms. This, in turn, results in difficulties in the formulation of realistic optimization objectives (Dunkel, 1970; Mendelssohn, 1978).
7 Reed (1980) handled this problem by including density dependency in a discrete-time model of an age-structured population.
8 A patient player is the one who discounts the future less heavily, in other words, the player with the lower discount rate.
9 In a competitive market model, there are so many interacting agents that the impact of the actions of one agent on the market can be assumed to be negligible. In an oligopolistic market model, however, there are few agents whose actions have significant impacts.
10 This part of the Soviet Union is now Russia.
11 Since this essay was first written, new work has been done on this issue and indications are that for reserves to hedge against uncertainty in a bioeconomic sense, net transfer rates must be "reasonably" high and reserve sizes must be large: large reserves provide good protection for the stock in the face of the uncertainty, while high transfer rates make the protected fish available for fishing after the shock has occurred (Sumaila, 1998a).
12 The feedback Nash equilibrium concept usually does not lend itself to numerical computation, except for two extreme cases (Carlson and Haurie, 1996): (i) the linear-quadratic game structure; (ii) the affine dynamics (see for instance, Breton *et al.*, 1986; Kamien and Schwartz, 1991; Binmore, 1982).

3 Cooperative and non-cooperative management when capital investment is malleable

1 With kind permission from Springer Science + Business Media: based on *Environmental & Resource Economics*, "Cooperative and non-cooperative exploitation of the Arcto-Norwegian cod stock in the Barents Sea," 10(2): 1997, 147–165, Sumaila, U.R.
2 One can think of T as a consortium of Norwegian, Russian and third party countries' trawler fleets, and C as the Norwegian coastal fleet.
3 Except where otherwise stated, cooperation is used in this chapter to mean cooperation with no predetermined catch shares.
4 It should be noted that here, unlike in the non-cooperative case, agents are concerned about the effects of changes in the stock level on their combined payoffs.
5 In this model, recruitment refers to the number of age zero fish that enter the habitat in each fishing period.
6 $\alpha = f'(0)$, is the number of recruits per unit weight of biomass "at zero."
7 Note that Norwegian data are applied in the analysis.
8 This gives a maximum biomass level of about 5 million tonnes when the model is run without fishing, that is, a maximum sustainable yield (MSY) of 2.5 million tonnes: assuming that the MSY stock level is one half of the pristine stock level. Note that this estimate is more conservative than the MSY stock level of about 3 million tonnes estimated by researchers at the Institute of Marine Research, Bergen (pers. comm.).
9 Note: Approx. NOK6.5 = US$1.
10 According to Fisheries Statistics 1989–1990, Tables 23 and 26, mature cod in 1989 got a price premium of 15% compared to juvenile cod: I thank an anonymous referee for pointing this out to me.
11 For the sake of scaling, units of fishing effort of 10 trawlers and 150 coastal vessels are used. The costs per fleet are calculated using cost data in Kjelby (1993): data related to the most cost effective vessels in T and C were used. These turned out to be vessels of size 13 and 21 m stationed in Nordland, Troms, and Finnmark, in the case of coastal fisheries, and factory trawlers in the case of trawlers. Note: Approx. NOK6.5 = US$1.
12 Powersim is a dynamic simulation software package developed by ModellData AS in Bergen, Norway (http://www.powersim.com/). The package has many powerful features, including the ability to process array variables.
13 The cooperative regime CPUEs compare favorably with the estimates of 12,320 and 790 tonnes for the two types of vessels reported in Kjelby (1993).
14 This can be done by introducing, say, oligopolistic markets, instead of the competitive markets assumed in the current version of the model.

4 Cooperative and non-cooperative management when capital investment is non-malleable

1 Based on *Marine Resource Economics*, 10(3): Sumaila, U.R. "Irreversible capital investment in a two-stage bimatrix fishery game model," 263–283, 1995, with permission from The MRE Foundation.
2 The cod-like fishes group includes the North-east Atlantic cod, Arcto-Norwegian haddock, whiting, saith, etc.
3 Data in tables E21–E51 in Lønnsomhetsundersokelser (1979–1990) were used for the calculations.
4 See Sumaila (1994) for the justifications for these simplifications.
5 Cost effectiveness is defined here in terms of least cost per kilogram of fish landed.
6 In a sense, one can argue that the game formulated herein is not a "pure" open loop strategy game. This is because although the fishing capacities are chosen once and for all, the capacity utilization is chosen in each period depending on the stock size.

7 Recall that the subscript denoting player is i = T, C.
8 In this model, recruitment refers to the number of age zero fish that enter the habitat in each fishing period.
9 $\alpha = f'(0)$, is the number of recruits per unit weight of biomass "at zero" or the population level.
10 Researchers at the Institute of Marine Research, Bergen, estimate the maximum sustainable yield (MSY) stock level to be about 3 million tonnes: with an assumption that the MSY stock level is one half of the pristine stock level, 6 million tonnes are obtained.
11 The price per kilogram of NOK 6.78 is taken from Table 22, Central Bureau of Statistics of Norway (1989–1990). The parameters ξ_i and ϕ_i are calculated using cost data in Lonnsomhetsundersokelser (1979–1990).
12 The 1992 stock size is estimated at 1.8 million tonnes (Ressursoversikt, 1993).
13 A practical way to view these variations is that the agents in this model have alternative uses for their vessel capacities, thereby making it possible for them to divert excess capacity in any given year to such uses.
14 Hannesson (1993a) looks more closely at the possible gains from allowing mobility of vessels between different stocks.
15 Note that these catches comprise both Norwegian and Russian landings, since the two are not differentiated in the model.
16 Recall that the elasticity of a function, f(x,y), with respect to x is defined as the percentage increase in f(x,y) resulting from a 1% increase in x.
17 Relative profitability is defined as discounted resource rent to T divided by discounted resource rent to C multiplied by 100.
18 The choice of this level will depend on both the depreciation rate and the difference between the cost of acquiring a new vessel and the disinvestment resale price of the vessels relative to the price of new vessels.

5 Strategic dynamic interaction: the case of Barents Sea fisheries

1 Based on *Marine Resource Economics*, 12, Sumaila, U.R. "Strategic dynamic interaction: The case of Barents Sea fisheries." 77–94, 1997a, with permission from The MRE Foundation.
2 Eide and Flaaten (1992) gave a comprehensive review of the ecosystem of the Barents Sea fisheries.
3 Contrast this with Flaaten and Armstrong (1991), where two variants of the cooperative (joint management) solution are discussed; one in which transfer payments are allowed and the other in which they are not. The sole ownership assumption here coincides with the "transfer payment" variant.
4 Clark and Kirkwood (1979) is one example where a similar formulation of the catch function is applied.
5 It should be noted, however, that the curve is concave only in the relevant range. In fact, it has a point of inflection near $B_{\text{prey}} = 0.05$.
6 Note that equation (5.6) enters the profit function of the cod owner.
7 Here, recruitment refers to the number of age zero fish that enter the habitat in each fishing period.
8 $\alpha = f'(0)$, is the number of recruits per unit weight of biomass "at zero" or the population level.
9 Moxnes (1992), however, is a study where recruitment functions are specified for both cod and capelin.
10 Strict concavity is ensured in the objective functional of players through the way cost functions are modeled.
11 In the separate management case, it is quite conceivable that the owners of the two fisheries may face different discount factors, for example, because they value future

benefits differently. Joint management, however, will imply that the same discount factor is applied for the two fisheries.

6 Cannibalism and the optimal sharing of the North-east Atlantic cod stock

1 With kind permission from Springer Science + Business Media: Based on *Journal of Bioeconomics*, "Cannibalism and the optimal sharing of the North-East Atlantic cod stock: A bioeconomic model," 2, 2000, 99–115, C.W. Armstrong and U.R. Sumaila.
2 The use of the word optimal in this text refers to situations that ensure maximum profits over a given period of time.
3 Coastal vessels use mainly passive fishing gear such as gillnets, hand lines and Danish seines. In addition, some longlines are employed.
4 There are historical, geographic, and biological reasons to account for the fact that Norway has a coastal small-scale fleet, while Russia does not. The Norwegian coastline is more hospitable towards both small vessels and fish resources than the northern Russian coastline. Also, the former Soviet government has prioritized industrial fisheries over the years. This left no room for small-scale inshore enterprises.
5 In actual fact, two different allocation rules have been implemented sequentially. Armstrong (1998) gives an account both of the first allocation rule which was implemented in 1990, and the somewhat improved allocation rule that has been implemented since 1994. It is this latter allocation rule that was applied in this chapter.
6 Eide (1997, p. 5) shows that the model gives a surprisingly good correspondence to historical catches and stock estimates from VPA runs' for the North-east Atlantic cod stock for a period of almost 30 years.
7 Note that r_1 is a somewhat modified intrinsic growth rate compared to the traditional use of the expression, as the stock growth as $x_1 \rightarrow 0$ is not just r_1 but rather $r_i + bx_2$.
8 This is a simplification as there is overlap of fishing, i.e. the coastal vessels catch some immature cod, and the trawlers catch some mature cod. Nonetheless, the two vessel groups do target different sections of the cod stock. Armstrong (1999) shows that in 1993 almost 60% of the trawl catch consisted of individuals less than 7 years of age; more than 70% of the coastal vessel catch consisted of individuals 7 years and older.
9 The use of this non-linear cost function allows us to avoid the often politically unacceptable fishing moratorium, which inevitably results from the bang-bang solution in an overexploited fishery. This is done without changing the costs substantially from the more common linear case.
10 Readers who are interested in the Powersim simulation program code may contact the author.
11 Eide (1997) estimates r_i, a_i and b simultaneously using equations (6.2), and minimizing the relative sum of squares. The inputs in the estimates are historic biomass and catch data for North-east Atlantic cod from 1962 to 1990.
12 In Fiskeribladet, 16. July 1997 (in Norwegian), a central Norwegian biologist is critical to fishing mature cod to reduce the predatory pressure on the immature cod.

7 Implications of implementing an ITQ management system for the Arcto-Norwegian cod stock

1 Based on Armstrong, Claire W. and Ussif Rashid Sumaila "Optimal allocation of TAC and the implications of implementing an ITQ management system for the North-east Arctic cod." Originally published in *Land Economics*, 77.3 (2001): 350–358. ©2001 by the Board of Regents of the University of Wisconsin System. Reproduced courtesy of the University of Wisconsin Press.

2 This is probably the first time a study is based on learning something about single species management from a multispecies perspective; the opposite is usually the case. I thank an anonymous referee for making me aware of this.
3 For allocational and regulation purposes, the Norwegian fisheries management divides the vessels that catch cod into two groups – trawlers and coastal vessels. The trawler group consists of all cod fishing trawlers, including factory vessels. The coastal vessel group is a more heterogeneous group of usually smaller size than the trawlers, which utilizes a number of different gear types.
4 Hannesson (1996) shows how, despite ITQs, one may get situations where concentration does not occur. That is, fish prices may increase (or costs may decrease) such that it is optimal to have less than the original quota share, hence sale of shares, and more partakers in the fishery may be the result. Matthiasson (1997) illustrates how local government involvement in the quota market, via subsidization of local vessel owners in order to secure job creation or preservation, also can avoid concentration of quota. Furthermore, less effective vessel firms find ways to circumvent the profitability consequences of the ITQ management, as a result of the initial allocation of fishing rights.
5 Another reason is the transboundary nature of the fishery, which will be commented upon later in the chapter.
6 This is not very likely since given the existing vessel structure with coastal vessels only present in Norway, a concentration of quotas in the hands of coastal vessel owners would require a large degree of across the border trading in quotas. That is, in order for the quotas to be concentrated in the hands of coastal vessel owners, Norwegian coastal vessel owners would have to buy quotas not only from Norwegian trawler vessel owners, but also from Russian trawler vessel owners. Alternatively, the Russians would have to invest in coastal vessels, which may not be likely given the current state of the Russian economy. However, to illustrate the possible effects of ITQs, it is necessary to study a concentration of quota within the coastal vessel group. For concentration of quotas in the trawler group, Norwegian trawlers need only buy rights from Norwegian coastal vessel owners, making across the border trading of quotas unnecessary.
7 The reasons for this difference between Russian and Norwegian fishing are many. For one, the Russian coastline and its fish resources are not as amicable to small coastal vessels as the Norwegian, and the Soviet industrialization did not give priority to small fishing units. In the past, general fishery policy in Norway tended to favor coastal vessel owners through subsidization. This may also have given Norwegian coastal vessels an additional reason for their continued existence.
8 In the past, the coastal vessel group has not succeeded in fishing its entire allocated share in some years. The remainder has then been transferred to the trawler group, resulting in higher trawler catch shares in those years.
9 Many of the trawler vessels are owned by processing plants. In order to pay the crews as little as possible, the plants pay the minimum price for the fish, taking the profits out elsewhere. This may partially explain the low trawler price. A sensitivity analysis where the trawler prices were increased by 15% gives the trawlers higher profits per tonne than the coastal vessels, for the years studied. The optimal trawler share, however, increased less than 4%.
10 This is a simplification as there is overlap of fishing, that is, the coastal vessels catch some immature cod, and the trawlers catch some mature cod. Nevertheless, Armstrong (1999) shows that in 1993 almost 60% of the trawler catch consisted of individuals less than seven years of age, while more than 70% of the coastal vessel catch consisted of individuals seven years and older.
11 NOK denotes the Norwegian Kroner. NOK 5.523 = US$1 as at 27 January 2013.
12 Since one can easily convert catch in a given period to effort level employed in that period, only the latter is reported.

13 It should be kept in mind that optimal economic data are used in these results. Doing the same analysis, but using data more in line with actual costs gives lower β values both in the optimal case ($\beta = 0.6$), as well as for the case where the catch share is closest to the trawl ladder ($\beta = 0.8$). Hence, the current management arrangement favors the trawlers, but not to the same degree that the optimal data seem to imply.
14 These values are used rather than 0 and 1, as the latter values result in "end point" computational problems. In addition, it may not be realistic to expect 100% concentration in one vessel type.

8 Marine protected area performance in a game-theoretic model of the fishery

1 Based on *Natural Resource Modeling*, 15(4), Sumaila, U.R., "Marine protected area performance in a model of the fishery" 439–451, 2002, with permission of The Natural Resource Modeling Association.
2 Thanks to Scott Farrow of Department of Economics, University of Maryland for providing this information.
3 Executive Order 13158, 26 May 2000, available at www.whitehouse.gov.
4 This function was chosen because recent biological studies have shown that it is more realistic than the Ricker recruitment function for species such as cod (Guénette and Pitcher, 1999).
5 Clearly catch costs may be affected by the MPA size, making allowance for longer travel distance. However, this would depend on the structure and positioning of the MPA, as well as the fisher's alternatives: issues that are beyond the scope of this chapter.
6 This is the minimum spawning biomass recommended to ensure the long-term sustainability of the North-east Atlantic cod (Nakken *et al.*, 1996).

9 Distributional and efficiency effects of marine protected areas

1 Based on Sumaila, Ussif Rashid and Claire W. Armstrong. "Distributional and efficiency effects of marine protected areas: A study of the North-East Atlantic cod fishery." Originally published in *Land Economics*, 82.3 (2006): 321–332. ©2006 by the Board of Regents of the University of Wisconsin System. Reproduced courtesy of the University of Wisconsin Press.
2 This is a more suitable model to use for the unmanaged scenario than open access, as limited entry, that is, regulated open access (Homans and Wilen, 1997) has been practiced for many years in the North-east Atlantic cod fishery. In later years, more stringent management has been put in place, and one could argue that a cooperative model better describes the fisheries outcome today.
3 An interesting extension of the present chapter would be to introduce some random shocks into the model.
4 Two of the northern Norwegian counties fishermen's organizations have suggested closing-off sections of the Barents Sea in order to protect juvenile cod.
5 This function is chosen because recent biological studies have shown that it is more realistic than the Ricker recruitment function for species such as cod (Guénette and Pitcher, 1999).
6 Hence, a reserve as presented here could be seen as an aggregation of a more extensive network of reserves spanning the many different areas that a migratory fish stock may traverse. If such a network sampled the entire area of the stock's life-cycle movement, it could, in theory, be designed such that it impeded in an equal fashion upon both vessel groups, hence reducing some of the distributional issues in this chapter. However, networks of reserves are usually designed with biological issues in

mind and could therefore exacerbate conflict between the two groups by closing one of the groups' historical fishing grounds.

7 In most marine reserve models so far, density dependent migration has been the rule. As Gell and Roberts (2003) point out, very little research has been done to test whether migration actually is density dependent. Attwood (2002) shows how the movement of galjoen out of a South African marine reserve was independent of fish densities. In vastly migratory species such as the North-Eeast Atlantic cod stock, density dependent migration does not seem like a natural assumption.

8 Fishing costs may be affected by the MPA size, making for longer travel distance. However, this would depend on structure and positioning of the MPA, as well as the fisher's alternatives; issues that are not studied here.

9 This is the spawning biomass seen as the long-term management goal for the North-east Atlantic cod (Anon., 2000a, b).

10 This ratio consists of all fishing of the North-east Atlantic cod stock; Russian, Norwegian, and other countries' trawler fishing versus the Norwegian coastal vessel fishing. Approximately 50% of the catch is taken by the Russian fleet which consists mainly of trawler vessels. This fleet structure is historic and a result of Soviet policies of industrialization.

10 Playing sequential games with Western Central Pacific tuna stocks

1 Based on Sumaila, U.R. and Bailey, M. (2011). Sequential fishing of Western Central Pacific Ocean tuna stocks. *Fisheries Centre Working Paper #2011-02*. University of British Columbia.

2 Here, recruitment refers to the number of age zero fish that enter the habitat in each fishing period.

3 $\chi = f'(0)$, is the number of recruits per unit weight of biomass "at zero."

11 Impact of management scenarios and fisheries gear selectivity on the potential economic gains from a fish stock

1 Based on Sumaila, U.R. (1999b). Impact of management scenarios and fishing gear selectivity on the potential economic gains from Namibian hake. CMI Working Paper 1999: 3.

2 The use of longliners to exploit hake is expected to increase with time, producing impacts on both the biology and economics of hake exploitation.

3 Note that the catchability of a fishing gear is defined as the share of the total stock being caught by one unit of fishing effort. On the other hand, the selectivity parameter of a fishing gear is the probability of the gear to retain fish of a particular age group.

4 Clearly, this is one of the assumptions in the current model that needs to be researched and improved upon in future applications of the model.

5 The rule consists of two steps. First, each player must receive his or her threat point payoffs. Second, the surplus over the sum of the threat point payoffs of all players is split equally between the players. The rationale for this sharing formula is that, to satisfy the individual rationality constraint (Binmore, 1982), players must be guaranteed their payoff under a non-cooperative regime, after which the surplus should be shared equally because each party to the cooperative agreement contributed equally to its success.

6 N$ denotes Namibian dollar. US$1 = N$8.97 on 27 January 2013.

12 Managing bluefin tuna in the Mediterranean Sea

1 Based on *Marine Policy*, 36(2), Sumaila, U.R. and L. Huang, "Managing bluefin tuna in the Mediterranean Sea," 502–511, 2012, with permission from Elsevier.

2 West Atlantic BFT breeds mostly in the Gulf of Mexico (Clay, 1991).
3 http://www.rthk.org.hk/rthk/news/englishnews/20110105/news_20110105_56_724679.htm. *RTHK*. 2011-01-05. Retrieved 2011-01-05.
4 Since 86.5% of the catch is caught with purse seine in 2006, it is reasonable to use only purse seine fishing costs.
5 The 48 contracting parties as of 2010 are United States, Japan, South Africa, Ghana, Canada, France, Brazil, Morocco, Republic of Korea, Cote d'Ivoire, Angola, Russia, Gabon, Cap-Vert, Uruguay, São Tomé and Principe, Venezuela, Guinea Ecuatorial, Guinee Rep, United Kingdom, Libya, China, Croatia, EU, Tunisie, Panama, Trinidad & Tobago, Namibia, Barbados, Honduras, Algerie, Mexico, Vanuatu, Iceland, Turkey, Philippines, Norway, Nicaragua, Guatemala, Senegal, Belize, Syria, St Vincent & the Grenadines, Nigeria, Egypt, Albania, Sierra Leone and Mauritania.
6 Spain has a system of licensing that limits vessel power and gear usage (Garza-Gil *et al.*, 1996), Syria licenses vessels based on approval by the fisheries department, and Turkey has a strict vessel and licensing system. There is some evidence that many other Mediterranean countries have licensing based effort controls but little official documentation has been found.
7 See http://ec.europa.eu/regional_policy/funds/prord/prords/prdsd_en.htm for more information. FIFG has recently been replaced by a new European Fisheries Fund (EFF) 2007–2013, established by EC Regulation 1198/2006.
8 MPA Global is a worldwide project for MPAs. Refer to http://www.mpaglobal.org/index.php?action=aboutus for more details.
9 See http://www.cites.org/eng/cop/index.shtml for detailed information.
10 In reality, no information about the quota selling price is currently found, but some data exist for other parameters. Thus, in this example, sensitivity analysis is carried out for quota selling prices.

Appendix: theoretical basis of the solution procedure

1 A game is of perfect information if the players always know everything that has happened so far in the game.

References

Alemany, F., Quintanilla, L., Velez-Belchí, P., García, A., Cortés, D., Rodríguez, J.M., Fernándeze Puelles, M.L., González-Pola, C., and López-Jurado, J.L. (2010). Characterization of the spawning habitat of Atlantic bluefin tuna and related species in the Balearic Sea (western Mediterranean). *Progress in Oceanography*, 86(1–2): 21–38.

Anderson, L.G. (1975a). Criteria for maximizing the economic yield of an internationally exploited fishery. *In* H.G. Knight (ed.) *The Future of International Fisheries Management*. St Paul, MN: West.

Anderson, L.G. (1975b). Optimum economic yield on an internationally utilized common property resource. *Fishery Bulletin*, 73: 51–56.

Anderson, L.G., and Sutinen, J.G. (1984). A survey of stochastic bioeconomics: methods and results. *Marine Resource Economics*, 1(2): 117–136.

Anon. (1990a, 1996, 1997, 2000a). Resource overview. Norwegian Fisheries Directorate, Bergen, Norway (in Norwegian).

Anon. (1990b, 1991, 1992, 1993). Annual profitability investigations for fishing vessels. Fiskeridirectoratet, Bergen, Norway (in Norwegian).

Anon. (2000b). Resources of the ocean. Havets Ressurser. Institute of Marine Research (in Norwegian).

Armstrong, C. (1998). Sharing a fish resource – bargaining theoretical analysis of an applied allocation rule. *Marine Policy*, 22: 119–134.

Armstrong, C. (1999). Sharing a resource – bioeconomic analysis of an applied allocation rule. *Environmental & Resource Economics*, 13: 75–94.

Armstrong, C., and Flaaten, O. (1991a). Non-cooperative and cooperative management of transboundary, interdependent natural resources. Department of Economics and Administration, University of Tromsø, Working Paper.

Armstrong, C., and Flaaten, O. (1991b). The optimal management of a transboundary fish resource: the Arcto-Norwegian cod stock. *In* R. Arnason and T. Bjørndal (eds) *Essays on the Economics of Migratory Fish stocks*. 137–152.

Armstrong, C., and Sumaila, U.R. (2000). Cannibalism and the optimal sharing of the North-east Atlantic cod stock: a bioeconomic model. *Journal of Bioeconomics*, 2(2): 99–115.

Armstrong, C., and Sumaila, U.R. (2001). Optimal allocation of TAC and the implications of implementing an ITQ management system for the North-east Atlantic cod. *Land Economics*, 77(3): 350–358.

Armstrong, C., Bergland, H., Eide, A., Flaaten, O., and Larsen, N.J. (1991). Torskefiskets Økonomi og regulering (The economics and management of the cod fishery). Rapport SR0509, FORUT, Tromsø (in Norwegian).

Arnason, R. (1991). On optimal harvesting of migratory species. *In* R. Arnason and T. Bjørndal (eds) *Essays on the Economics of Migratory Fish Stocks*. 21–40.

Arnason, R., Magnusson, G., and Agnarson, S. (2000). The Norwegian spring-spawning herring fishery: a stylized game model. *Marine Resource Economics*, 15: 293–319.

Asche, F., Salvanes, K.G., and Steen, F. (1997). The curse of uncertainty in a fishery. Centre for Fisheries Economics Report No. 37, Foundation for Research in Economics and Business Administration, Bergen, Norway.

Attwood, C.G. (2002). Spatial and temporal dynamics of an exploited reef-fish population. PhD dissertation, University of Cape Town, South Africa.

Bailey, M., Sumaila, U.R., and Lindroos, M. (2010). Application of game theory to fisheries over three decades. *Fisheries Research*, 102: 1–8.

BBC News. http://news.bbc.co.uk/2/hi/science/nature/7040011.stm, released on 17 October 2007.

Bertsekas, D.P., and Tsitsiklis, J.N. (1989). *Parallel and Distributed Computation.* New York: Prentice-Hall.

Beverton, R.J.H., and Holt, S.J. (1957). On the dynamics of exploited fish population. *Fisheries Investigation Series*, 2(19). London: Ministry of Agriculture.

Binmore, K. (1982). *Fun and Games: A text on game theory.* London: Chancellor Press.

Binmore, K. and P. Dasgupta, ed., (1986a). 'Game Theory: A Survey' in Economic Organisations as games. Basil Blackwell, Oxford.

Binmore, K. and P. Dasgupta, ed., (1986b). 'Nash Bargaining Theory: An Introduction' in *The Economics of Bargaining*. Basil Blackwell, Oxford.

Bjørndal, T. (1992). Management of fisheries as a common property resource. Paper presented at the *XXI Conference of the International Association of Agricultural Economics*. Tokyo, 22–29 August.

Bjørndal, T., and Munro, G. (2012). *The Economics and Management of World Fisheries*. New York: Oxford University Press.

Block, B.A., Dewar, H., Blackwell, S.B., Williams, T.D., Prince, E.D., Farwell, C.J., Boustany, A., Teo, S.L.H., Seitz, A., Walliand, A., and Fudge, D. (2001). Migratory movements, depth preferences, and thermal biology of Atlantic Bluefin Tuna. *Science*, 293(5533): 1310–1314.

Bohnsack, J.A. (1990). The potential of marine fishery reserves for reef fish management in the US Southern Atlantic. Miami: NOAA Tech. Memo NMFS-SEFC-261. 40 pp.

Bohnsack, J.A. (1993). Marine reserves. They enhance fisheries, reduce conflicts, and protect resources. *Oceanus*, Fall: 63–71.

Bonfil, R., Munro, G., Sumaila, U.R., Valtysson, H., Wright, M., Pitcher, T., Preikshot, D., Haggan, N., and Pauly, D. (1998). Impacts of distant water fleets: an ecological, economic and social assessment. *In The Footprints of Distant Water Fleets on World Fisheries*. Endangered Seas Campaign. Godalming: WWF International. 11–111.

Borch, K. (1962). Application of game theory to some problems in automobile insurance. *The Astin Bulletin*, 2(2): 208–221.

Boyce, J.R. (1992). Individual transferable quotas and productions externalities in a fishery. *Natural Resource Modeling*, 5(4): 385–408.

Braithwaite, R.B. (1955). *Theory of Games as a Tool for the Moral Philosopher.* Cambridge: Cambridge University Press.

Brasao, A., Duarte, C.C., and Cunha-e-S, M.A. (2000). Managing the Northern Atlantic bluefin tuna fisheries: the stability of the UN Fish Stock Agreement Solution. *Marine Resource Economics*, 15: 341–360.

Bregazzi, R.M. (2007). The plunder of bluefin tuna in the Mediterranean and East Atlantic during 2006 and 2007. ATRT, SL. RMB & Partners.

Breton, M., Filar, J.A., Haurie, A., and Schultz, T.A. (1986). On the computation of equilibria in discounted stochastic dynamic games. *In* T. Basar (ed.) *Dynamic Games*

and Applications in Economics. Lecture Notes Economics and Math. Systems 265. Berlin: Springer-Verlag.
Byrknes, A. (1996). *Run-time U ser's Guide and Reference Manual*. Powersim. 57 pp.
Carlson, D.A., and Haurie, A. (1996). A turnpike theory for infinite-horizon open-loop competitive processes. *SIAM Journal of Control and Optimization*, 34(4): 1405–1419.
Carroll, M.T., Anderson, J.L., and Martines-Garmendia, J. (2001). Pricing US North Atlantic bluefin tuna and implications for management. *Agribusiness*, 17(2): 243–254.
Cavazutti, E., and Flåm, S.D. (1992). Evolution to selected Nash equilibria. *In* E. Giannessi (ed.) *Nonsmooth Optimisation Methods and Application*. London: Gordon and Beach.
Central Bureau of Statistics Norway (1989–1990). *Fishery Statistics*. Oslo: The Bureau.
Charles, A.T. (1983). Optimal fisheries investment: comparative dynamics for a deterministic seasonal fishery. *Canadian Journal of Fisheries and Aquatic Sciences*, 40: 2069–2079.
Charles, A.T., and Munro, G.R. (1985). Irreversible investment and optimal fisheries management: a stochastic analysis. *Marine Resource Economics*, 10(3): 247–264.
Christy, F.T. Jr, and Scott, A.D. (1965). *The Common Wealth in Ocean Fisheries*. Baltimore, MD: John Hopkins University Press.
Chu, C. (2009). Thirty years later: the global growth of ITQs and their influence on stock status in marine fisheries. *Fish and Fisheries*, 10(2): 217–230.
Clark, C.W. (1980). Restricted access to common-property fishery resources: a game theoretic analysis. In Liu. P. (ed.) *Dynamic Optimisation and Mathematical Economics*, Plenum Press.
Clark, C.W. (1990). *Mathematical Bioeconomics: The optimal management of renewable resources*. New York: Wiley Interscience.
Clark, C.W. (1996). Marine reserves and the precautionary management of fisheries. *Ecological Applications*, 6(2): 369–370.
Clark, C.W. (2010). *Mathematical Bioeconomics: The mathematics of conservation*, 3rd edn. New York: Wiley Interscience.
Clark, C.W., and Kirkwood, G.P. (1979). Bioeconomic model of the Gulf of Carpentaria prawn fishery. *Journal of the Fisheries Research Board of Canada*, 36: 1304–1312.
Clark, C.W., and Munro, G.R. (1975). The economics of fishing and modern capital theory: a simplified approach. *Journal of Environmental Economics and Management*, 2: 92–106.
Clark, C.W., Clarke, F.H., and Munro, G.R. (1979). The optimal exploitation of renewable resource stocks: problems of irreversible investment. *Econometrica*, 47: 25–49.
Clay, D. (1991). Atlantic bluefin tuna (*Thunnus thynnus*): a review. *In* R.B. Deriso and W.H. Bayliff (eds) *World Meeting on Stock Assessment of Bluefin Tunas: Strengths and weaknesses*. Inter-American Tropical Tuna Commission Special Report No. 7. 357 pp.
Clemhout, S., and Wan Jr, Y. (1985). Cartelization conserves endangered species. *In* G. Feichtinger (ed.) *Optimal Control Theory and Economic Analysis*, Vol. 2. Amsterdam: North-Holland. 549–568.
Concerted Action (2006). Economic Performance of Selected European Fishing Fleets, Annual Report 2005. Concerted Action on Promotion of Common Methods for Economic Assessment of EU Fisheries under EC contract FISH/2005/12.
Concerted Action (2007). Economic Performance of Selected European Fishing Fleets 2008. The Potential Economic Impact on Selected Fleet Segments of TACs proposed by ACFM and reviewed by SGRST for 2008 (EIAA-model calculations).
Conrad, J.M., and Clark, C.W. (1987). *Natural Resource Economics: Notes and problems*. Cambridge: Cambridge University Press.

Copes, P. (1981). Rational resource management and institutional constraints: The case of the fishery. *In* John A. Butlin (ed.) *Economics and Resource Policy*. London: Longman. 113–128.

Copes, P. (1986). A critical review of the individual quota as a device in fisheries management. *Land Economics*, 62, 278–291.

Costello, C., Gaines, S.D., and Lynham, J. (2008). Can catch shares prevent fisheries collapse? *Science*, 321(5896): 1678–1681.

Council Regulation (EC). No 1451/2001 of 28 June 2001.

Council Regulation (EC). No 2287/2003 of 19 December 2003.

Council Regulation (EC). No 27/2005 of 22 December 2004.

Cullen, R. (1996). Fisheries management: lessons about "tradable quotas" from New Zealand. *Choices*, Third Quarter: 29–31.

Cury, P. (1994). Obstinate nature: An ecology of individuals. Thoughts on reproductive behaviour and biodiversity. *Canadian Journal of Fisheries and Aquatic Science*, 51: 1664–1673.

Datta, M., and Mirman, L.J. (1999). Externalities, market power, and resource extraction. *Journal of Environmental Economics and Management*, 37(3): 233–255.

Davis, G.E. (1981). On the role of underwater parks and sanctuaries in the management of coastal resources in the southeastern United States. *Environmental Conservation*, 8: 67–70.

Desse, J., and Desse-Berset, N. (1994). Stratégies de pêche au 8ème millénaire: les poissons de Cap Andreas Kastros (Chypre). *In* A. Le Brun (ed.) *Fouilles récentes à Khirokitia*, Editions Recherche sur les Civilisations, Paris. 335–360.

Dewees, C.M. (1989). Assessment of the implementation of individual transferable quotas in New Zealand's inshore fishery. *North American Journal of Fisheries Management*, 9: 131–139.

Digernes, T. (1980). Matematisk Modellering Av Fiskebestander (Mathematic modeling of fish stocks. Institutt for Atomenergi, Report No. ESS-27 (in Norwegian).

Dockner, E., Feichtinger, G., and Mehlmann, A. (1989). Non-cooperative solutions for differential game model of fishery. *Journal of Economic Dynamics and Control*, 13: 1–20.

Duarte, C.C., Brasao, A., and Pintassilgo, P. (2000). Management of the Northern Atlantic bluefin tuna: an application of C-games. *Marine Resource Economics*, 15: 21–36.

Dudley, N., and Waugh, G. (1980). Exploitation of a single-cohort fishery under risk: a simulation-optimization approach. *Journal of Environmental Economics and Management*, 7: 234–255.

Dunkel, G.M. (1970). Maximum sustainable yields. *SIAM Journal of Applied Mathematics*, 19: 1629–1640.

Dyck, A.J., and Sumaila, U.R. (2010). Economic impact of ocean fish populations in the global fishery. *Journal of Bioeconomics*, 12(3): 227–243.

Eide, A. (1997). Stock variations due to symbiotic behaviour – an attempt to understand the fluctuations in the cod fisheries in the Barents Sea. *In* McDonald and McAleer (eds) *MODSIM 97, International Congress on Modelling and Simulation Proceedings*. http://tidley.nfh.uit.no/arne/stock_variations.pdf. 1599–1604.

Eide, A., and Flaaten, O. (1992). Bioeconomic modelling of the Barents Sea Fisheries. Paper presented at the *6th Conference of the International Institute of Fisheries Economics and Trade*, Session No. 3 Part 1. Paris.

Eikeland, P.O. (1993). Distributional aspects of multispecies management. The Barents Sea Large Marine Ecosystem. *Marine Policy*, 17(4): 256–271.

Fischer, R.D., and Mirman, L.J. (1992). Strategic dynamic interaction: fish wars. *Journal of Dynamics and Control*, 16: 267–287.

Fischer, R.D., and Mirman, L.J. (1996). The compleat fish wars: biological and dynamic interactions. *Journal of Environmental Economics and Management*, 30(1): 34–42.

Flaaten, O. (1988). *The Economics of Multispecies Harvesting – Theory and application to the Barents Sea fisheries*. Studies in Contemporary Economics. Berlin: Springer-Verlag.

Flaaten, O., and Armstrong, C. (1991). Non-cooperative and cooperative management of transboundary, interdependent natural resources. Department of Economics and Admininstration, University of Tromsø Working Paper.

Flåm, S.D. (1993). Path to constrained Nash equilibria. *Applied Mathematics and Optimisation*, 27: 275–289.

Flåm, S.D. (1994). Kor Mykje er dei Norske Sildefiskeria Verdt? (What is the value of the Norwegian Herring fishery?) *Sosialøkonomen*, 48(4): 2–5 (in Norwegian).

Frank, D.M., and Sarkar, S. (2010). Group decisions in biodiversity conservation: implications from game theory. *PLoS ONE*, 5(5): e10688.

Fromentin, J.-M. (2006). ICCAT Manual Chapter 2.1.5: Atlantic Bluefin Tuna. ICCAT.

Fromentin, J.-M., and Powers, J.E. (2005). Atlantic bluefin tuna: population dynamics, ecology, fisheries and management. *Fish and Fisheries*, 6: 281–306.

Fromentin, J.-M., and Ravier, C. (2005). The East Atlantic and Mediterranean bluefin tuna stock: looking for sustainability in a context of large uncertainties and strong political pressures. *Bulletin of Marine Science*, 76: 353–362.

Fudenberg D. and Tirole (1991), Game Theory, MIT Press.

Gaitsgory, V., and Nitzan, S. (1994). A folk theorem for dynamic games. *Journal of Mathematical Economics*, 23: 167–178.

Garza-Gil, D., Iglesias-Malvido, C., Suris-Regueiroand, J.C., and Varela-Lafuente, M.M. (1996). The Spanish case regarding fishing regulation. *Marine Policy*, 20(3): 249–259.

Geen, G., and Nayar, M. (1988). Individual transferable quotas in the southern bluefin tuna fishery: an economic appraisal. *Marine Resource Economics*, 5: 365–387.

Gell, F.R., and Roberts, C.M. (2003). *The Fishery Effects of Marine Reserves and Fishery Closures*. Washington DC: World Wildlife Fund.

Gordon, H.S. (1954). Economic theory of a common-property resource: the fishery. *Journal of Political Economy*, 62: 124–142.

Grafton, R.Q. (1996). Individual transferable quotas: theory and practice. *Reviews in Fish Biology and Fisheries*, 6: 5–20.

Grafton, R.Q., Hannesson, R., Shallard, B., Sykes, D., and Terry, J. (2006). The economics of allocation in Tuna Regional Fisheries Management Organizations (RFMOs). Australian National University, Economics and Environment Network Working Paper EEN0612.

Guénette, S., and Pitcher, T.J. (1999). An age-structured model showing the benefits of marine reserves against overexploitation. *Fisheries Research*, 39(3): 295–303.

Guth, W., Schmittberger, K. Schwarze, B. (1982). An experimental analysis of ultimatum bargaining. *Journal of Economic Behaviour and Organisation*, 3: 367–88.

Halpern, B.S. (2003). The impact of marine reserves: do reserves work and does reserve size matter? *Ecological Applications*, 13: S117–S137.

Halpern, B.S., and Warner, R.R. (2002). Marine reserves have rapid and lasting effects. *Ecology Letters*, 5: 361–366.

Hamre, J., and Tjelmeland, S. (1982). *Sustainable Yield Estimate of the Barents Sea Capelin Stock*. Copenhagen: ICES.

Hamukuaya, H. (1994). Research to determine the biomass of hake. *Namibian Brief: Focus on Fisheries and Research*, 18: 73–74.

Hannesson, R. (1978). Tralfiske eller skreifiske? Om optimal beskatning av den norske-arktiske torkbestanden. (Trawl fishing or coastal fishing for mature cod? On the optimal management of the Arcto-Norwegian cod stock. Universitetet I Bergen, Økonomiske skrifter nr. 12. Bergen (in Norwegian).

Hannesson, R. (1983). Optimal harvesting of ecologically interdependent fish species. *Journal of Environmental Economics and Management*, 10: 329–345.

Hannesson, R. (1993a). *Bioeconomic Analysis of Fisheries*. London: Fishing News Books.

Hannesson, R. (1993b). Optimal fishing capacity and international transfer of excess allowable catches. *In* J.R. Nielsen and T. Helgason (eds) *Multispecies Fisheries Management.* Proceedings of a symposium in Hiirtshals. 22–23 May. 44 pp.

Hannesson, R. (1995). Sequential fishing: cooperative and non-cooperative equilibria. *Natural Resource Modeling*, 9(1): 51–59.

Hannesson, R. (1996). Long-term industrial equilibrium in an ITQ managed fishery. *Environmental & Resource Economics*, 8: 63–74.

Hannesson, R. (1998). Marine reserves: what would they accomplish? *Marine Resource Economics*, 13(3): 159–170.

Hannesson, R. (2011). Game theory and fisheries. *Annual Review of Resource Economics*, 3: 181–202.

Hannesson, R., Herrick, S., and Field, J. (2009). Ecological and economic considerations in the conservation and management of the Pacific sardine (*Sardinops sagax*). *Canadian Journal of Fisheries and Aquatic Sciences*, 66: 859–868.

Hardin, G. (1968). The tragedy of the commons. *Science*, 162: 1243–1248.

Harsanyi, J. (1982). Solutions for some bargaining games under the Harsanyi Selten solution theory. *Mathematical Social Sciences*, 3: 179–91 and 259–89.

Haurie, A., and Roche, M. (1993). Turnpikes and computation of piecewise open-loop equilibria in stochastic differential games. *Journal of Economic Dynamics and Control*, 18: 317–344.

Haurie, A., and Tolwinski, B. (1990). Cooperative equilibria in discounted stochastic sequential games. *Journal of Optimisation and Applications*, 64(3): 511–535.

Hawkshaw, R.S., Hawkshaw, S., and Sumaila, U.R. (2012). The tragedy of the 'tragedy of the commons': why coining too good a phrase can be dangerous. *Sustainability*, 4(11): 3141–3150.

Herrick, S.F., Norton, J.G., Hannesson, R., Sumaila, U.R., Ahmed, M., and Pena-Torres, J. (2009). Global production and economics. *In* D.M. Checkley, J. Alheit, Y. Oozeki and C. Roy (eds) *Climate Change and Small Pelagic Fish.* Cambridge: Cambridge University Press. 256–274.

Holland, D.S. (2000). A bioeconomic model of marine sanctuaries on Georges Bank. *Canadian Journal of Fisheries and Aquatic Science*, 57: 1307–1319.

Holland, D.S. (1998). The use of year around closed areas for the management of New England trawl fisheries. PhD dissertation, University of Rhode Island, USA.

Holland, D.S., and Brazee, R.J. (1996). Marine reserves for fisheries management. *Marine Resource Economics*, 3(11): 157–171.

Holling, C.S. (1965). The functional response of predators to prey density and its role in mimicry and population regulations. *Memoirs of the Entomology Society of Canada*, 45: 1–60.

Homans, F.R., and Wilen, J. (1997). A model of regulated open access resource use. *Journal of Environmental Economics and Management*, 32(1): 1–21.

Horwood, J.W., and Whittle, P. (1986). Optimal control in the neighbourhood of an optimal equilibrium with examples from fisheries models. *IMA Journal of Mathematics Applied in Medicine and Biology*, 3: 129–142.

ICCAT (1994). Report for biennial period, 1992–93 Part II (1993) Vol. 2. SCRS. Madrid, Spain.

ICCAT (1995). Report for biennial period, 1994–95 Part I (1994) Vol. 2. SCRS. Madrid, Spain.

ICCAT (1996). Report for biennial period, 1994–95 Part II (1995) Vol. 2. SCRS. Madrid, Spain.

ICCAT (2005). Report for biennial period, 2004–05 Part I (2004) Vol. 2. SCRS. Madrid, Spain.

ICCAT (2006). Report for biennial period, 2004–05 Part II (2005) Vol. 2. SCRS. Madrid, Spain.

ICCAT (2007). Report for biennial period, 2006–07 Part I (2006) Vol. 2. SCRS. Madrid, Spain.

ICCAT (2008a). Report for biennial period, 2006–07 Part II (2007) Vol. 2. SCRS. Madrid, Spain.

ICCAT (2008b). Report of the 2008 Atlantic Bluefin Tuna Stock Assessment Session. Madrid.

ICCAT (2009). Report for biennial period, 2008–09 Part I (2008) Vol. 2. SCRS. Madrid, Spain.

ICCAT (2010). Report for biennial period, 2008–09 Part II (2009) Vol. 2. SCRS. Madrid, Spain.

ICES (1992). Report on the Arctic fisheries working group. Copenhagen.

ICES (1996). Report on the Arctic fisheries working group. Copenhagen.

Institute of Marine Research. (1993). *Ressursoversikt*, Særnummer 1, Bergen (in Norwegian).

Institute of Marine Research (1994). Ressursoversikt: fisken og havet (Fish and the sea). Særnummer 1, Bergen (in Norwegian).

Ishimura, G., Herrick, S., and Sumaila, U.R. (2012). Fishing games under climate variability: trans boundary management of Pacific sardine in the California Current System. *Environmental Economics and Policy*, DOI: 10.1007/s10018-012-0048-0.

Ishimura, G., Herrick, S., and Sumaila, U.R. (2013). Stability of cooperative management of the Pacific sardine fishery under climate variability. *Marine Policy*, 39: 333–340.

Jakobsen, T. (1993). Management of Northeast Arctic-cod – past, present and future? *Proceedings of Symposium on Management Strategies for Exploited Fish Populations*, October, University of Alaska, Anchorage.

Kaitala, V.T. (1985). *Game Theory Models of Dynamic Bargaining and Contracting in Fisheries Management*. Institute of Mathematics, Helsinki University of Technology.

Kaitala, V.T. (1986). Game theory models in fisheries management: a survey. *In* T. Basar (ed.) *Dynamic Games and Applications in Economics*. Berlin: Springer-Verlag.

Kaitala, V. and Lindroos, M. (1998). Sharing the benefits of cooperation in high seas fisheries: a characteristic-function game approach. *Natural Resource Modeling*, 11: 275–299.

Kaitala, V.T., and Munro, G.R. (1993). The management of high seas fisheries. *Marine Resource Economics*, 8: 313–329.

Kaitala, V.T., and Munro, G.R. (1995). The economic management of high seas fishery resources: some game theoretic aspects. *In* C. Carraro and J.A. Filar (eds) *Control and Game Theoretic Models of the Environment*. 299–318.

Kaitala, V.T., and Pohjola, M. (1988). Optimal recovery of a shared stock: a differential game with efficient memory equilibria. *Natural Resource Modelling*, 3: 91–119.

Kamien, M.I., and Schwartz, N.L. (1991). *Dynamic Optimisation*. Amsterdam: North Holland.

Kelleher, G. (1999). Guidelines for marine protected areas. World Conservation Union, Gland, Switzerland.

Kennedy, J.O.S., and Watkins, J.W. (1986). Time-dependent quotas for the southern bluefin tuna fishery. *Marine Resource Economics*, 2(4): 293–313.

Kjelby, T. (1993). Det norske torskefiskeriet som nasjonalformue, SEFOS N OTAT 90 (in Norwegian).

Klieve, H., and MacAulay, T.G. (1993). A game theory analysis of management strategies for the southern bluefin tuna industry. *Australian Journal of Agricultural Economics*, 37: 17–32.

Krawczyk, J.B., and Tolwinski, B. (1991). A cooperative solution for the three-agent Southern Bluefin Tuna Management problem. Paper presented at the *15th IFIP Conference on System Modelling and Optimisation*. Zurich, Switzerland, 2–6 September.

Kronbak, L.G., and Lindroos, M. (2007). Sharing rules and stability in coalition games with externalities. *Marine Resource Economics*, 22: 137–154.

Kyndland, F.E., and Prescott, E.C. (1977). Rules rather than discretion: the inconsistency of optimal plans. *Journal of Political Economy*, 85(3): 471–491.

Larsen, R.G., and Isaksen, B. (1993). Size and selectivity of rigid sorting grids in bottom trawls for Atlantic cod (*Gadus morhua*) and haddock (*Melanogrammus aeglefinus*). *ICES Marine Science Symposium*, 196: 178–182.

Lauck, T. (1996). Uncertainty in fisheries management. *In* D.V. Gordon and G.R. Munro (eds) *Fisheries and Uncertainty: A Precautionary Approach to Resource Management*. Calgary: University of Calgary Press. 91–105.

Lauck, T., Clark, C.W., Mangel, M., and Munro, G.R. (1998). Implementing the precautionary principle in fisheries management through marine reserves. *Ecological Applications*, 8(1): S72–S78.

Laukkanen, M. (2003). Cooperative and non-cooperative harvesting in a stochastic sequential fishery. *Journal of Environment Economics and Management*, 45(2): 454–473.

Leslie, P.H. (1945). On the use of matrices in certain population mathematics. *Biometrika*, 35: 183–212.

Levhari, D. and Mirman, L.J. (1980). The great fish war: an example using a dynamic Cournot-Nash solution. *Bell Journal of Economics*, 11: 322–334.

Lewis, E.G. (1942). On the generation and growth of populations. *Sankhya*, 6: 93–96.

Lewontin, R.C. (1961). Evolution and the theory of games. *Journal of Theoretical Biology*, 1: 382–403.

Lindroos, M. (2004). Sharing the benefits of cooperation in the Norwegian spring-spawning herring fishery. *International Game Theory Review*, 6(1): 35–53.

Lindroos, M., Kaitala, V., and Kronbak, L.G. (2007). Coalition games in fisheries economics. *In* T. Bjørndal, D.V. Gordon, R. Arnason and U.R. Sumaila (eds) *Advances in Fisheries Economics: Festschrift in honour of Professor Gordon Munro*. Oxford: Blackwell. 184–195.

Lotka, A.J. (1925). *Elements of Physical Biology*. (Reissued as *Elements of Mathematical Biology*, by Dover, 1956). Baltimore, MD: Williams and Wilkins.

MacCall, A.D. (1990). *Dynamic Geography of Marine Fish Populations*. Washington Sea Grant Program, Seattle, WA: University of Washington Press. 153 pp.

McInerney, J.P. (1981). Natural resource economics: the basic analytical principles. *In* John A. Butlin (ed.) *Economics and Resource Policy*. London: Longman. 30–58.

McKelvey, R.W. (1997). Game-theoretic insights into the international management of fisheries. *Natural Resource Modeling*, 10: 129–171.

Mackinson, S., Sumaila, U.R., and Pitcher, T.J. (1997). Bioeconomics and catchability: fish and fishers behavior during stock collapse. *Fisheries Research*, 31(1–2): 11–17.

McRae, D., and Munro, G.R. (1989). Coastal state 'rights' within the 200-mile Exclusive Economic Zone. *In* P.A. Neher, R. Arnason and N. Mollett (eds) *Rights Based Fishing*. 97–111.

Madani, K. and Dinar, A. (2012). Cooperative institutions for sustainable common pool resource management: application to groundwater. *Water Resources Research*, 48(9): W09553.

Man, A., Law, R., and Polunin, N.V.C. (1995). Role of marine reserves in recruitment to reef fisheries: a metapopulation model. *Biological Conservation*, 71: 197–204.

Matthiasson, T. (1997). Local government and the Icelandic ITQ market. *Marine Resource Economics*, 8: 63–74.

May, R.M., Beddington, J.R., Clark, C.W., Holt, S.J., and Laws, R.M. (1979). Management of multispecies fisheries. *Science*, 205: 267–277.

Mendelssohn, R. (1978). Optimal harvesting strategies for stochastic single species, multiage class models. *Mathematical Bioscience*, 41: 159–174.

Mertens, J.F. and Zamir, S. (1985). Formulations and analysis for games with incomplete information. International Journal of Game Theory, 14.

Miller, C., Munro, G., Sumaila, U. R., and Cheung, W. W. L. (2013). Governing marine fisheries in a changing climate: a game theoretic perspective. *Canadian Journal of Agriculture Economics*, forthcoming.

Miller, K.A., and Munro, G.R. (2004). Climate and cooperation: a new perspective on the management of shared fish stocks. *Marine Resource Economics*, 19: 367–393.

Ministry of Fisheries and Marine Resources (1996). Fishery Statistics. Summary Statistical Report. Namibia.

Miyake, P.M. (1992). Tuna Catches by ICCAT Non-member Countries. International Commission for the Conservation of Atlantic Tunas. NMFS. Draft Environmental Assessment for the US Western Atlantic Bluefin Tuna Fishery. National Marine Fisheries Services, 31 pp.

Miyake, P.M., De la Serna, J.M., Di Natale, A., Farrugia, A., Katavic, I., Miyabe, N., and Ticina, V. (2003). General review of bluefin tuna farming in the Mediterranean area. *Collective Volume of Scientific Papers ICCAT*, 55(1): 114–124.

Moorsom, R. (1994). An analysis of the fishing capacity of the Namibian fleet in quota-limited fisheries: a methodological summary. NEPRU Working Paper no. 39.

Moxnes, E. (1992). Multispecies management under uncertainty. SNF (Centre for Research in Economics and Business Administration). Report No. 95. Norwegian School of Business and Administration, Bergen. 34 pp.

MSBN. http://www.msnbc.msn.com/id/5428979/, released on 19 July 2004.

Munro, G. (1979). The optimal management of transboundary renewable resources. *Canadian Journal of Economics*, 12(8): 355–376.

Munro, G. (1990). The optimal management of transboundary fisheries: game theoretic considerations. *Natural Resource Modeling*, 4: 403–426.

Munro, G. (1991a). The management of transboundary fishery resources: a theoretical overview. *In* R. Arnason and T. Bjørndal (eds) *Essays on the Economics of Migratory Fish Stocks*. 7–20.

Munro, G. (1991b). The management of migratory fish resources in the pacific. *In* R. Arnason and T. Bjørndal (eds) *Essays on the Management of Migratory Fish Stocks*. 85–106.

Munro, G. (1996). Approaches to the economics of the management of high seas fishery resources. *In* D.V. Gordon and G. Munro (eds) *Fisheries and Uncertainty: A precautionary approach to resource management*. Calgary: University of Calgary Press. 161–180.

Munro, G. (1998). *The Management of High Seas Fisheries and the United Nations Conference on Straddling Fish Stocks and Highly Migratory Fish Stocks: A review*. Vancouver: University of British Columbia.

Munro, G., Houtte, A.V., and Willmann, R. (2004). *The Conservation and Management of Shared Fish Stocks: Legal and Economic Aspects*. Rome: Food and Agriculture Organization of the United Nations.

Myerson, R. (1984). An introduction to game theory. Discussion Paper No. 623. Northwestern University.

Nakken, O., Sandberg, P., and Steinshamn, S.I. (1996). Reference points for optimal fish stock management. *Marine Policy*, 20(6): 447–462.

Nasar, S. (2002). *A Beautiful Mind: Genius and schizophrenia in the life of John Nash*. New York: Faber and Faber.

Nash, J.F. (1950a). The bargaining problem. *Econometrica*, 18: 155–162.

Nash, J.F. (1950b). Equilibrium points in n-person games. *Proceedings of the National Academy of Science USA*, 36: 48–49.

Nash, J.F. (1951). Non-cooperative games. *Annals of Mathematics*, 54: 286–295.

Nash, J.F. (1953). Two-person cooperative games. *Econometrica*, 21: 128–140.

von Neumann, J. and Morgenstern, O. (1944). Theory of Games and Economic Behaviour. Princeton. Princeton University Press.

NMFS (2010). Supplemental Environmental Assessment, Regulatory Impact Review, and Final Regulatory Flexibility Analysis for the Final 2010 Atlantic Bluefin Tuna Quota Specifications. US Department of Commerce, NOAA, National Marine Fisheries Service, Office of Sustainable Fisheries, Highly Migratory Species Management Division; April.

OCEANA and MarViva (2008). Bluefin Tuna Larval Survey. 2008 Oceana-MarViva Mediterranean Project.

Ostrom, E. (1990). *Governing the Commons: The Evolution of Institutions for Collective Action*. Cambridge: Cambridge University Press.

Pallares, P., Soto, M., Die, D.J., Gaertner, D., Mosquiera, I., and Kell, L. (2005). The development of an operational model and simulation procedure for testing uncertainties in the Atlantic bigeye (*Thunnus obesus*) stock assessment. *ICCAT Collection of Scientific Papers*, 57(2): 162–176.

Pauly, D., Christensen, V., Guénette, S., Pitcher, T.J., Sumaila, U.R., Walters, C.J., Watson, R., and Zeller, D. (2002). Towards sustainability in world fisheries. *Nature*, 418: 689–695.

Payne, A.I.L., O'Brien, C.M., and Rogers, S.I. (eds) (2004). *Management of Shared Fish Stocks*. Abingdon: Wiley-Blackwell. 384 pp.

Pezzey, J.C.V., Roberts, C.M., and Urdal, B.T. (2000). A simple bioeconomic model of marine reserves. *Ecological Economics*, 33(1): 77–91.

Pikitch, E.K., Rountos, K.J., Essington, T.E., Santora, C., Pauly, D., Watson, R., Sumaila, U.R., Boersma, P.D., Boyd, I., Conover, D.O., Cury, P., Heppell, S.S., Houde, E.D., Mangel, M., Plagányi, É., Sainsbury, K., Steneck, R.S., Geers, T.M., Gownaris, N., and Munch, S.B. (2012). The global contribution of forage fish to marine fisheries and ecosystems. *Fish and Fisheries*, DOI: 10.1111/faf.12004.

Pintassilgo, P. (2003). A coalition approach to the management of high seas fisheries in the presence of externalities. *Natural Resource Modeling*, 16(2): 175–196.

Pitcher, T.J., Watson, R., Haggan, N., Guènette, S., Kennish, R., Sumaila, U.R., Cook, D., Wilson, K., and Leung, A. (2000). Marine reserves and the restoration of fisheries and marine ecosystems in the South China Sea. *Bulletin of Marine Science*, 66: 543–566.

Polacheck, T. (1990). Year round closed areas as a management tool. *Natural Resource Modeling*, 4: 327–354.

Pontecorvo, G., Wilkinson, M., Anderson, R., and Holdowsky, M. (1980). Contribution of the Ocean Sector to the United States Economy. *Science*, 208: 1000–1006.

Porch, C.E. (2005). The sustainability of Western Atlantic Bluefin Tuna: a warm blooded fish in a hot blooded fishery. *Bulletin of Marine Science*, 76: 363–384.

Powers, J.E. (1992). Bluefin tuna stock trajectories under alternative catch histories. Memo to Dr William J. Fox, Jr, Director, National Marine Fisheries Service; 9 January.

Punt, A.E. (1988). Model selection for the dynamics of South African hake resources. MSc thesis, University of Cape Town, South Africa. Report of the Benguela Ecology Program of South Africa. 15: [vii]+395 pp.

Punt, A.E. and Butterworth, D.S. (1991). On an approach for comparing the implications of alternative fish stock assessment, with application to the stock of cape hake Merluccius spp. off northern Namibia. *South African Journal of Marine Science*, 10: 219–240.

Quirk, J.P. and Smith, V.L. (1977). Dynamic economic models of fishing. *In* A.D. Scott (ed.) *Economics of Fisheries Management, A symposium*. University of British Columbia, Institute of Animal Research and Ecology.

Reed, W.J. (1980). Optimum age-specific harvesting in a nonlinear population model. *Biometrics*, 36: 579–593.

Reid, C., Vakurepe, R., and Campbell, H. (2003). Tuna prices and fishing costs for bioeconomic modelling of the Western and central pacific tuna fisheries. Technical Report, ACIAR Project No. ASEM/2001/036.

Reithe, S. (2002). Marine reserves and quotas as fisheries management tools: when does marine reserve creation pay? *Proceedings of the Eleventh Biennial Conference of the International Institute of Fisheries Economics and Trade (IIFET)*. Wellington, New Zealand.

Renton, A. (2008). How the world's oceans are running out of fish. *The Observer*, released 11 May.

Ricker, W.E. (1954). Stock and recruitment. *Journal of Fisheries Research Board of Canada*, 11: 559–623.

Rodwell, L.D., Barbier, E.B., Roberts, C.M., and McClanahan, T.R. (2003). The importance of habitat quality for marine reserve – fishery linkages. *Canadian Journal of Fisheries and Aquatic Science*, 60(2): 171–181.

Rosen, J.B. (1965). Existence and uniqueness of equilibrium points for concave n-person games. *Econometrica*, 33: 520–534.

Salm, R.V., and Clark, J.R. (1989). Marine and coastal protected areas: a guide for planners and managers. Gland: IUCN.

Sanchirico, J.N., and Wilen, J.E. (1999). Bioeconomics of spatial exploitation in a patchy environment. *Journal of Environmental Economics and Management*, 37(2): 129–150.

Sanchirico, J.N., and Wilen, J.E. (2001). A bioeconomic model of reserve creation. *Journal of Environmental Economics and Management*, 42(3): 129–150.

Schaefer, M.B. (1957). Some considerations of population dynamics and economics in relation to the management of marine fisheries. *Journal of Fisheries Research Board of Canada*, 14: 669–681.

Shapley, L.S. and Shubik, M. (1954). A method for evaluating the distribution of power in a committee system. *American Political Science Review*, 48: 787–792.

Shubik, M. (1962). Incentives, decentralised Control, the assignment of joint costs and internal pricing. Management Science 8, 325–343.

Silvert, W., and Smith, W.R. (1977). Optimal exploitation of a multispecies community. *Mathematical Bioscience*, 33: 121–134.

Slade, M.E. (1995). Empirical games: the oligopoly case. *Canadian Journal of Economics*, 18(2): 368–402.

Smith, V.L. (1969). On models of commercial fishing. *Journal of Political Economy*, 77: 181–198.

Spulber, D.F. (1985). The mulitcohort fishery under uncertainty. *Marine Resource Economics*, 1(3): 265–282.

Sumaila, U.R. (1994). A two-agent model for the Arcto-Norwegian cod. Department of Economics, University of Bergen, Working Paper No. 0694. 18 pp.

Sumaila, U.R. (1995). Irreversible capital investment in a two-stage bimatrix game model. *Marine Resource Economics*, 10(3): 163–183.

Sumaila, U.R. (1996). Essays on the application of game theory in the management of natural resources. Dissertations in Economics No. 12. Department of Economics, University of Bergen, Norway.

Sumaila, U.R. (1997a). Strategic dynamic interaction: the case of Barents Sea fisheries. *Marine Resource Economics*. 12: 77–94.

Sumaila, U.R. (1997b). Cooperative and non-cooperative exploitation of the Arcto-Norwegian cod stock in the Barents Sea. *Environmental & Resource Economics*, 10(2): 147–165.

Sumaila, U.R. (1998a). Protected marine reserves as hedge against uncertainty: an economist's perspective. *In* T.J. Pitcher, D. Pauly and P. Hart (eds) *Reinventing Fisheries Management*.London: Chapman and Hall. 301–309.

Sumaila, U.R. (1998b). Protected marine reserves as fisheries management tools: a bioeconomic analysis. *Fisheries Research*, 37(1–3): 287–296.

Sumaila, U.R. (1999a). A review of game theoretic models of fishing. *Marine Policy*, 23(1): 1–10.

Sumaila, U.R. (1999b). Impact of management scenarios and fishing gear selectivity on the potential economic gains from Namibian hake. CMI Working Paper 1999: 3.

Sumaila, U.R. (2000). Fish as vehicle for economic development in Namibia. *Forum for Development Studies*, 2: 295–315.

Sumaila, U.R. (2002). Marine protected area performance in a model of the fishery. *Natural Resource Modeling*, 15(4): 439–451.

Sumaila, U.R. (2012a). Seas, oceans and fisheries: a challenge for good governance. *The Round Table: The Commonwealth Journal of International Affairs*, 101(2): 157–166.

Sumaila, U.R. (2012b). Overfishing: call to split fisheries at home and abroad. *Nature* 481, DOI:10.1038/481265c.

Sumaila, U.R. and Bailey, M. (2011). Sequential fishing of Western Central Pacific Ocean tuna stocks. Fisheries Centre Working Paper 2011-02. University of British Columbia, Vancouver.

Sumaila, U.R., and Charles, A. (2002). Economic models of marine protected areas: an introduction. *Natural Resource Modeling*, 15(3): 261–272.

Sumaila, U.R., and Huang, L. (2012). Managing bluefin tuna in the Mediterranean Sea. *Marine Policy*, 36: 502–511.

Sumaila, U.R., Guénette, S., Alder, J., and Chuenpagdee, R. (2000). Addressing ecosystem effects of fishing using marine protected areas. *ICES Journal of Marine Science*, 57: 752–760.

Sumaila, U.R., Alder, J., and Keith, H. (2006). Global scope and economics of illegal fishing. *Marine Policy*, 30(6): 696–703.

Sumaila, U.R., Khan, A., Dyck, A., Watson, R., Munro, G., Tyedmers, P., and Pauly, D. (2010). A bottom-up re-estimation of global fisheries subsidies. *Journal of Bioeconomics*, 12(3): 201–225.

Sumaila, U.R., Cheung, W.W.L., Lam, V.W.Y., Pauly, D., and Herrick, S. (2011). Climate change impacts on the biophysics and economics of world fisheries. *Nature Climate Change*, 1: 449–456.

Sundaram R. (1989). Nash equilibrium in a class of symmetric dynamic games: An existence theorem, Journal of Economic Theory 47,153 177.

Susannah F.L. (2008). http://www.scienceline.org/2008/03/env-locke-tuna/ released 7 March.

Teh, L., and Sumaila, U.R. (2011). Contribution of marine fisheries to worldwide employment. *Fish and Fisheries*, DOI: 10.1111/j.1467-2979.2011.00450.x.

Tjelmeland, S. (1990). MULTSIMP: a simplified multispecies model for the Barents Sea. MULTSPEC support papers, Institute of Marine Research, Bergen.

Tomkins, J., and Butlin, J.A. (1975). A theoretical and empirical approach to fisheries economics. *Journal of Agricultural Economics*, 26: 105–125.

United Nations (1994). Draft Agreement for the Implementation of the Provisions of the United Nations Convention on the Law of the Sea of 10 December 1982 Relating to the Conservation and Management of Straddling Fish Stocks and Highly Migratory Fish Stocks. UN Document A. Conference 164:22.

Vislie, J. (1987). On the optimal management of transboundary renewable resources: a comment on Munro's paper. *Canadian Journal of Economics*, 20: 870–875.

Volterra, V. (1928). Variations and fluctuations of the number of individuals in animal species living together. *ICES Journal du Conseil*, 3: 1–51.

Wallis, O.L. (1971). Establishing underwater parks worldwide. *Transactions of the North American Wildlife and Natural Resource Conference*, 36: 97–115.

Walters, C.J. (2000). Impacts of dispersal, ecological interactions and fishing effort dynamics on efficacy of marine protected areas: how large should protected areas be? *Bulletin of Marine Science*, 66: 754–757.

Walters, C.J., and Hilborn, R. (1978). Adaptive control of fishing systems. *Journal of the Fisheries Research Board of Canada*, 33: 145–159.

Wilen, J.E. and Homans, F.R. (1997). Unraveling rent losses in modern fisheries: production, market, or regulatory inefficiencies. *In* E.K. Pikitch, D.D. Huppert and M.P. Sissenwine (eds) *Proceedings of the Symposium Global Trends, Fisheries Management*, Seattle, Washington. American Fisheries Society.

Worm, B., Hilborn, R., Baum, J.K., Branch, T.A., Collie, J.S., Costello, C., Fogarty, M.J., Fulton, E.A., Hutchings, J.A., Jensen, O.P., Lotze, H.K., Mace, P.M., McClanahan, T.R., Minto, C., Palumbi, S.R., Parma, A.M., Ricard, D., Rosenberg, A.A., Watson, R., and Zeller, D. (2009). Rebuilding global fisheries. *Science*, 325: 578–585.

WWF (2004). Tuna farming in the Mediterranean: the bluefin tuna stock at stake. WWF.

WWF (2006). The plunder of bluefin tuna in the Mediterranean and East Atlantic in 2004 and 2005. WWF Mediterranean Programme Office.

Wysokinski, A. (1986). The living marine resources of the Southeast Atlantic. FAO Fisheries Technical Paper, 178 Revision 1.

Index

Note: Page references in *italics* indicate figures and page references in **bold** indicate tables.